YOUR HUMAN GEOGRAPHY DISSERTATION

Designing, Doing, Delivering

Kimberley Peters

Los Angeles | London | New Delhi
Singapore | Washington DC | Melbourne

Los Angeles | London | New Delhi
Singapore | Washington DC | Melbourne

SAGE Publications Ltd
1 Oliver's Yard
55 City Road
London EC1Y 1SP

SAGE Publications Inc.
2455 Teller Road
Thousand Oaks, California 91320

SAGE Publications India Pvt Ltd
B 1/I 1 Mohan Cooperative Industrial Area
Mathura Road
New Delhi 110 044

SAGE Publications Asia-Pacific Pte Ltd
3 Church Street
#10-04 Samsung Hub
Singapore 049483

Editor: Robert Rojek
Editorial assistant: Matthew Oldfield
Production editor: Katherine Haw
Copyeditor: Catja Pafort
Indexer: Judith Lavender
Marketing manager: Sally Ransom
Cover design: Stephanie Guyaz
Typeset by: C&M Digitals (P) Ltd, Chennai, India
Printed by CPI Group (UK) Ltd, Croydon, CR0 4YY

First published 2017

Library of Congress Control Number: 2016944499

British Library Cataloguing in Publication data

A catalogue record for this book is available from the British Library

ISBN 978-1-4462-9518-2
ISBN 978-1-4462-9520-5 (pbk)

'This excellent new text guides students carefully, intelligently and sympathetically through the process of doing a human geography dissertation. It offers grounded advice - from the question of what a dissertation is, to the mechanics of data analysis - which will be indispensable for students researching the full diversity of topics covered by contemporary human geography. The insights, advice and reflections from both previous students and academic staff who currently teach human geography add valuable insights that will both reassure students and help them avoid making common mistakes.'

Peter Kraftl, Professor of Human Geography, University of Birmingham

'This book will be an invaluable read for all Human Geography dissertation students. It conveys the excitement and possibilities of Human Geography research, whilst also alerting the reader to its challenges and pitfalls. This is certainly not a generic "how to do your dissertation" textbook; instead it engages with Human Geography as a discipline and the role of the dissertation student as a producer of geographic knowledge. The book's clear sections on designing, doing and delivering your dissertation, have useful examples, include input from the author's students themselves, making this an accessible and comprehensive text.'

Katie Willis, Professor of Human Geography, Royal Holloway, University of London

'Kim Peters has written a much needed book that will be of great value to Geography students undertaking what is often the most challenging part of their degree, the dissertation. As a Geography lecturer I have often wished that a book such as this existed. *Your Human Geography Dissertation* goes way beyond a standard examination of the pros and cons of different research methods, covering a range of topics from the identification of dissertation subjects and the development of research questions through gathering data and writing up. It is a readable and highly accessible text full of helpful detail, practical advice and useful examples. Thank you Kim!'

Jo Little, Professor in Geography, University of Exeter

CONTENTS

ABOUT THE AUTHOR

Kim is a Lecturer in Human Geography at the University of Liverpool. She has previously held lecturing posts at Aberystwyth University and the University of Sheffield, following the completion of a PhD at Royal Holloway, University of London in 2012. Before becoming an academic Kim worked as a transport planner, civil servant and as a sales advisor in a London bike store. In her spare time she enjoys road cycling and visiting the coast. Kim's research focuses on the social, cultural and political organisation and use of maritime space and the geographies of mobilities. She has published widely in this area, including the co-edited books, *Waterworlds: Human Geographies of the Ocean* (Ashgate, 2014); *The Mobilities of Ships* (Routledge, 2015); and *Carceral Mobilities* (Routledge, 2017). She teaches in this area as well as more broadly on research methods and dissertation training. This is Kim's first textbook.

OTHER CONTRIBUTORS

Jonathan Duckett is a Human Geography PhD student at Loughborough University. His current research focuses on Scottish youth citizenship and national identity in relation to the cultural and political events of the Glasgow 2014 Commonwealth Games and the Scottish Independence Referendum.

Cordelia Freeman is a Teaching Associate in the School of Geography at the University of Nottingham. Her PhD was an examination of the history of violence on the Chile–Peru border and her work continues to explore themes of international diplomacy, military violence, and the biopolitics of health in the Latin American borderlands. Cordelia teaches on a number of political, historical, and cultural geography modules as well as research methods.

Amy Jones studies Human Geography at Swansea University, UK. Amy's doctoral research focuses on the physical act of walking the Wales Coast Path, investigating the ways in which experiences of the path are understood, felt and sensed through bodily actions and the ways in which performances of walking shape the people and places involved.

Sam Saville holds a BA in Human Geography, and MSc in Architecture: Advanced Environmental and Energy Studies. She has worked as an energy advisor (at the Centre for Sustainable Energy) and as a researcher and tutor at the Centre for Alternative Technology and as a visiting lecturer at Chester University. Currently she is a research assistant for the ERC-funded Global-Rural project and is completing a PhD in Human Geography at Aberystwyth University.

Robert Sheargold completed a BA in Human Geography at Aberystwyth University and a MA in Cultural Geography at Royal Holloway, University of London. He is currently working in London. He is interested in the relationship between mental well-being, product design and user experience research.

Emma Spence is a PhD student in the School of Geography and Planning at Cardiff University. Emma's research focuses upon elite and superrich mobility in the context of the luxury superyachting industry. She has published articles on this topic in the journals *Area* and *Mobilities*.

Rachael Squire is a Lecturer in Human Geography at Royal Holloway University. Her research interests centre on the geopolitics of undersea space with a specific focus on undersea habitats and US Navy experimental diving during the Cold War. More broadly, Rachael is interested in the intersections between extreme environments and the body, subterranean geopolitics, and ideas pertaining to 'territorial volume'.

Will Wright has recently completed a PhD in the Department of Geography at the University of Sheffield exploring the ongoing social and cultural legacies of the 2004 Indian Ocean tsunami in Sri Lanka. Will's broader research interests include postcolonial theory and the politics of knowledge production, social and cultural geographies of the sea, and critical tourism and development studies. He has also taught at undergraduate and postgraduate levels at the University of Sheffield.

ACKNOWLEDGEMENTS

This book has taken an incredibly long time to write. As an early-career academic, when I started out, I was confident in my ability to get the job done quickly. But as more experienced scholars told me (and they were right) it would take me longer than I imagined. My own expectations for what I hoped and wanted for the book held me back as I struggled to 'get it right'. I couldn't find the right voice, the right tone. I couldn't, at times, find the right words. And I wanted it to be right. This book has been a real passion – something for my own students, and something for the students of my colleagues. I wanted it to be right, for them. I hope that the following pages can guide, reassure and inspire. Any omissions, errors or shortcomings are my entirely own.

In trying to 'get it right' I have a number of people to thank for their influence, involvement and patience in the writing of this book. The first academic book I read cover-to-cover was Tim Cresswell's *In Place/Out of Place*. That text enlivened the geographer in me. Tim has been an important influence, both in terms of supervising my research but also in acting as an inspiration for how I hoped I might be able to someday write. I hope that this book goes at least a small way in meeting that aspiration.

I have also been fortunate enough to have benefited from working with a wonderful and supportive group of colleagues at Aberystwyth University who have watched this book develop from its inception: Liz Gagen, Jesse Heley, Laura Jones, Rhys Jones, Rhys Dafydd Jones, Mitch Rose, Marc Welsh, Mark Whitehead and Mike Woods. Particular thanks go to Gareth Hoskins who introduced me to the literature on writing practice used in Chapter 11, and Peter Merriman, who has been a significant source of encouragement throughout the writing process. I am also grateful to Andy Hardy and Rachel Smedley for their careful proof-reading of sections of Chapters 7 and 10 and to Will Andrews, Greg Thomas and the PhD cohort who helped me to develop new approaches to methods teaching. It was my teaching at Aberystwyth that provided the idea and context for this book and I will always be grateful for my time working in such a vibrant department.

I would also like to thank those from the wider academy who have assisted in helping me reach the finish line with this book. Firstly, my thanks go to Jon Anderson and Philip Crang who separately supervised my human geography dissertations, but together taught me what a dissertation is, and should be! My particular thanks also go to Jen Dickinson, Mark Holton, Innes Keighren, Sarah Mills, Jeanette Clarkin-Phillips, Sophie Wynne-Jones, Catherine Cottrell-Studemeyer, Andy Davies and Kevin Grove who, at various stages, have each offered a listening ear and a willing engagement with the project. I am also grateful for the broader academic support received as I completed the manuscript, with thanks to Jo Little, Peter Jackson, Katie Willis, Mike Brown, Peter Kraftl, Hilary Geoghegan, Peter Adey, Dominique Moran and Philip Steinberg. I would also like to express my thanks to the external reviewers who read this book at the proposal stage and as I completed draft chapters. Your feedback has been invaluable. Special thanks go to Jennifer Turner for providing the cover image for the book and for assisting me with the collation and formatting of images used in the text. Further thanks go to my colleagues at the University of Liverpool, and to the those at the University of Waikato, who enabled me the time and space to finally finish the manuscript.

I am also grateful to the contributors – those who have recently completed their degrees, current PhD students, and early-career scholars – in offering their time to provide recent graduate guidance. Being placed so close to the research process, these reflections have elevated many sections of the book with thoughtful and considered insights. I am also thankful for my students who – for the past 6 years – have continually amazed me with their ideas, their passion for geography and for pushing me to think of ways to make teaching practice more interesting!

Importantly, I would like to thank Robert Rojek at SAGE for his absolute patience and for putting his trust in this project from the moment he received my first email. Significant thanks must also go to Matthew Oldfield for his regular emails, swift responses to my numerous queries, and his words of encouragement which kept me (more or less) on track. My appreciation also goes to those who have been central to the production and marketing of the book and the development of the Companion Website, including Katherine Haw, Catja Pafort, Sally Ransom and Chloe Statham, and all those behind the scenes who have helped to make this book better than I could have ever hoped.

And finally, to Jennifer, who I owe the greatest thanks. You were there at the very beginning. You were there at the end. Your unwavering belief and constant support has sustained me throughout.

Kimberley Peters, Leicester, 2016.

LIST OF FIGURES

GRADUATE GUIDANCE

ABOUT THE COMPANION WEBSITE

As you read the book, don't forget to Go Online! and visit the companion website at **http://study.sagepub.com/yourhumangeography**.

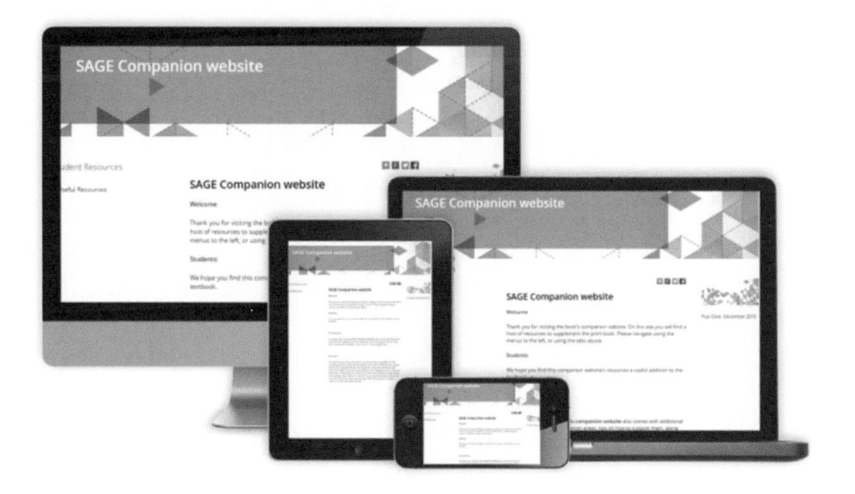

On the site, you'll find lots of helpful information and support, including:

- **Videos** of leading lecturers offering their top dissertation tips
- **Links** to useful free SAGE journal articles
- Great **research resources** including timetables, budgets and referencing guides
- **'Graduate Guidance'** reflections on previous dissertation experiences
- **Links** to useful websites
- **Take home messages** for each step in the dissertation process

1

YOUR HUMAN GEOGRAPHY DISSERTATION: AN INTRODUCTION

CHAPTER MAP

- Becoming a geographer
- What is a dissertation?
- Why do a human geography dissertation?
- Starting out: Top tips
- How to use this book

Becoming a geographer

If you are opening the pages of this book, it probably means you are at one stage or another of completing a degree in the subject of Geography. By now, through various modules and courses, you will certainly know what geography is, and what geographers do. The word 'geography' is a complex one that has taken on diverse meanings at different times (and for geographers with differing interests). However, it translates most readily to mean 'earth-writing' (Barnes and Duncan, 1992). The job of a geographer, very crudely then, is to write about the world. It further follows that the role of the *human* geographer is to describe and make sense of the relations between people, space and place, putting pen to paper in this respect. Human geographers are engaged with this task of 'authoring the world' from a variety of perspectives. Some geographers write about the changing environmental shape of our world for societies, others about the significance of socio-cultural phenomena: identity, citizenship, consumption, mobility, materiality (the list could go on). Some may engage with the demographic patterns that emerge in specific spaces over particular times. Others are concerned with the historical significance of the past on the present, whilst some explore how politics impacts contemporary understandings and engagements with space and place. The array of subjects, objects, persons and perspectives the human geographer can write about are wide-ranging.

However, in order to write about the world, we have to claim to know something about it. To know something about the world, we have to ask questions about the phenomena that shape it, impact it and change it. Therefore, there is a crucial connection between the role of the geographer as 'earth-writer' and the task of **research** (see Information Box 1.1). Indeed, in writing about the world, human geographers are not simply communicating ideas, they are *producing* knowledge. As Barnes and Duncan (1992) have noted, when we write about the world, we are also making the world; we are literally 'writing worlds'. Accordingly, the task of writing cannot be separated from the task of formulating knowledge; and the task of formulating knowledge cannot be detached from research. It is only from asking questions about the world that we can develop answers that reveal something we did not know or fully understand; which we can then commit to writing.

It is not just human geographers who engage with research. Within universities, academics in a number of fields – from the sciences to social sciences, arts and humanities – ask cutting-edge questions about varying topics and subjects in order to progress our understandings of them. In short, universities are not merely involved in teaching students about the world, they also conduct the research that produces the knowledge that is, in turn, taught. As Johnston and Sidaway explain,

> [m]ost disciplines are established in universities to teach a body of knowledge ... Their existence indicates a demand for instruction in that material by people who are an expert in it. Being an expert almost invariably means playing some part in producing and reproducing that body of knowledge. (2004: 2)

In other words, universities (and the experts who work there) don't just relate ideas and knowledge to us from the pages of a book; they write the book. Universities then, are sites of knowledge production.

INFORMATION BOX 1.1
WHAT IS RESEARCH?

To research is to investigate a subject, a topic, or a place. It is an endeavour that allows us to examine something that we might know little about or wish to know more about. Importantly it relies on us having a sense of what has been researched before. Researchers do not seek to address questions that have already been answered. They aim to consider new angles on current or existing problems, or, more ambitiously, to tackle completely new questions. Research is not 'ad-hoc', spontaneous, or random. It requires thought. Research is a systematic, careful and diligent study in order to test ideas, verify information and gather opinions to ultimately reach conclusions. Research is a process through which we are able to say something about our studies, that contributes to a body of knowledge. Moreover, research does not just add to our understanding, but might also act to actively make a change to a society, community, place or issue.

What is exciting is that it is not only experts who are involved in research and writing about the world. As part of your undergraduate studies you will have the opportunity to conduct your own research as part of a final year project or dissertation. This is arguably where you make the transition from being a *student* of geography to a *producer* of geography. In other words, it is during this process that you become a geographer. Whilst this might seem a daunting prospect, it needn't be if you consider why you chose to study geography in the first place. Most students pick the subject because they have an interest in the world, its people and its places. In effect, the dissertation is a realisation of what most students want to do anyway – investigate and study a portion of the world and find out more about it.

Geography is a relevant and contemporary subject. It isn't about abstract phenomena, but the world we are actually living in. The dissertation is your opportunity to delve into, study and research something about the world that you find interesting, adding (in however small a way) to a pool of existing knowledge. This book is concerned with how you become a geographer – partaking in the activity that we human geographers are involved in on a daily basis – that is, research. In what follows we consider what a dissertation is, why it's a worthwhile task, and consider some top tips for getting started.

What is a dissertation?

Arguably, the more we know about something the easier it is to understand. When completing any assessment we have the best chance of success by knowing what is expected of us (Kneale, 2011). So what is a dissertation, and, importantly, what sets it apart from other assignments you might do? The human geography dissertation can be defined in the following ways:

An extended piece of work

The dissertation is probably the longest piece of academic writing you will complete during your undergraduate degree. Extended projects and dissertations can range anywhere from 5,000 to 15,000 words (depending on your department or school).

As such, they offer an opportunity for you to display deep learning; that is your in-depth engagement with, and understanding of, a topic. It can often be overwhelming to think of the length of the dissertation at the start. However, knowing it is a lengthy endeavour can be helpful. Getting to grips with the expected word count early on makes it appear manageable. For example, dissertations are normally split into chapters – each chapter is typically the length of a standard essay that you are (by this point) used to writing. So splitting it down into these kind of chunks can help in the completion of the project (see also Chapter 11).

Moreover, a human geography dissertation is also lengthy in the period of time you have to complete it. Dissertations are typically researched and written over 9–12 months, covering the course of the second to third year of your degree (or third to fourth year for students studying in Scotland and those completing sandwich studies). Geography students tend to have unique timescales for their projects compared to other disciplines because of the nature of the research conducted and the need to collect data. Dissertations almost always straddle a period between differing years of a degree, providing the summer for collecting data and analysing findings. You have an ample portion of time for the dissertation (so remember you have time on your side) just as long as you are organised (you can download a timetable in the 'Starting out' section).

A piece of work requiring research

Talking of data collection, most dissertation projects can be differentiated from a standard essay in that you are not just considering and critiquing existing ideas from the literature in relation to a question set by a course tutor. You are formulating your own question(s) and completing your own research to provide answers. Dissertations almost always rely on the collection of data. They should therefore present that data as part of the finished product. Every dissertation should have empirics (that is, material from your investigations). That material could be anything from the analysis of a text (a book, a film, a website); interviews conducted with relevant persons; an ethnographic study where you situate yourself within a field of interest; a questionnaire survey; an investigation of an archive of historical records, or a 'triangulation' of methods (see Chapters 7, 8 and 9). Thus, it is paramount that your dissertation does not only reflect upon the current literature surrounding the topic you are interested in, but also offers some data towards our understanding of that topic.

A piece of work that makes a contribution to a topic area

It can be easy to get hung up about the issue of originality when it comes to a dissertation. It is often said that a dissertation must be original – featuring a brand new study of a novel question not previously answered. This is a myth. Certainly, a dissertation should have *elements* of originality. But it is worth remembering what you are trying to achieve in a dissertation project. A dissertation is limited to a specific number of words, over a set time period (with a clear deadline) and it is only you who will be conducting the research (often academics have a team, meaning the scale of their research can be much broader). Thus, a dissertation does not have to achieve anything earth-shatteringly new. Rather, a solid dissertation is one that contributes to an existing topic area. What might make your project original and new is that you apply existing questions and concerns to a new case study area or community; or you look at a place, a group of people, or a phenomenon from a different angle using an alternative set of theories or hypotheses. However, this doesn't mean you should lack ambition. The most important aspect of making a contribution lies in knowing what human geography research is already out there. In grasping the context to your field of study you can be best placed to make a sound, relevant contribution (see Clifford et al., 2010: 7; and Chapter 4 of this book).

An independent piece of work

Although most dissertations are set within a specific module and you will gain guidance throughout (normally via one-on-one supervision) the task of a dissertation is, in one sense, a test of how well you work independently; how well you can plan a course of study; manage your time, and produce something on your own. Staff will rarely badger you about the project. It is up to you to decide upon the topic, the question, the techniques for gathering data and the structure of the written report. You will need to be proactive and responsible for your project (Burkill and Burley, 1996). Whilst potentially daunting at first, this is, in many ways, the most exciting thing about the dissertation. It is a piece of work that you take ownership of. You can decide what aspect of human geography is of interest to you, and what you want to know more about.

A piece of work that requires you to demonstrate a range of skills

Degree programmes (in the UK) are subject to 'benchmarking standards' set out by the Quality Assurance Agency for Higher Education (QAA). Benchmarking standards are useful – they lay out the levels of achievement in various skills that you should gain by the end of your Geography degree (and they are well worth a look: see Information Box 1.2 for some highlights). Within QAA standards the dissertation is noted as an essential part of becoming a geographer. A dissertation involves demonstrating key skills in organisation, planning and task management; in formulating questions and selecting appropriate means for answering them, and in analysing and making sense of data. It is a key component to developing your skills and gaining new ones, which are essential to your academic ability and your prospects for future employment.

INFORMATION BOX 1.2
THE SKILLS OF A GEOGRAPHER

The QAA note that geography students should have the following attributes at the end of their degree. Students should be able to illustrate, evaluate and critically reflect upon:

- the issues involved in applying research **design** and execution skills within the specific context of **field-based** research
- the diversity of techniques and approaches involved in **collecting** geographical information
- specialised techniques and approaches involved in **analysing** geographical information
- specialised techniques and approaches involved in **presenting** geographical knowledge and information

(Source: taken from the QAA benchmarking statement 2014)

So we now know what a human geography dissertation is. However, every institution differs slightly in respect of what the dissertation is worth (in degree credits or weighting), the word count, the time you may take to complete it, the help and supervision you will receive and the exact skill range you should display. Therefore, it is essential – before you begin – that you familiarise yourself with your own institutional guidelines and regulations for completing the dissertation. Almost all schools and departments will have a handbook to help students get to grips with what the dissertation is and what is expected of you. Take time to find it, and read it. As we know, there's no better way to succeed than being prepared.

TASK 1.1

Find your departmental guidelines for the dissertation/extended project assignment. Common places to look might include your online learning environment. If it hasn't already been introduced, you could ask an appropriate staff member such as your personal tutor. Look over the regulations and make yourself familiar with the process in your department. It is particularly helpful to look over guidelines relating to what kind of **supervision** you can expect; **key deadlines** (for drafts and final submission); the **marking criteria** (so you know early on how you are being assessed); and any specific details about **formatting** that will save time by being employed at the start. Make a note of any questions or queries you have so you can ask the relevant person.

Students of mine have also found it helpful to speak to peers in the year above who are currently completing their dissertation, or Masters students who have already finished a research project. You may know students through societies or departmental activities, or it is possible your tutor may be able to put you in touch with some of their own final year students. Although personal experiences vary greatly, it is worth asking what challenges your peers have faced, and what information they wished they'd had at the start.

By considering the dissertation early on and arming yourself with as much information as possible, you can reduce the chance of any unexpected factors arising as you progress. Importantly, you can also limit any feelings of anxiety by being 'in the know'.

Why do a human geography dissertation?

Whilst we have explored what a dissertation is, you might still be asking why you need to complete one. Although it will likely be the case that you will be required to complete a dissertation project as part of your degree, there are various reasons why it is a beneficial – and even enjoyable – assessment to complete.

Inquisitiveness

One of the main reasons that students want to undertake a dissertation is because it provides an opportunity to study, in depth, a topic of real interest. There are few opportunities in a degree programme to simply 'run' with a topic that has grabbed your attention. In some senses a dissertation is truly a self-indulgent enterprise. We can follow our curiosity (Phillips and Johns, 2012). Yet the dissertation is also a chance to give back. Our inquisitiveness might have impacts. We might be able to help those people, communities, sites or places we work with, to better understand a current concern, emerging issue or longstanding problem. Indeed, the QAA note that an important element of completing a dissertation is in 'developing a sense of place, awareness of difference, and tolerance for others' (2014: n.p). In other words, it is about not just studying the world, but engaging with it in a sensitive, thoughtful and critical way.

Personal achievement

A dissertation is no small task. Developing a solid research plan, designing a framework for study, conducting research, making sense of it and writing it up often comes with a sense of personal achievement. However, whilst completing a dissertation is quite an accomplishment – in and of itself – many students also feel a sense of achievement in what it helps them to understand. This might come from adding to your own knowledge base, for example, via your ability to grasp a complex theory, use a new method or employ a specific analytic technique. However, a sense of achievement might also emerge from what your project can 'give back'. Sometimes, our research leads us to work with businesses, institutions, local community groups, charities, and so on, and, in turn, our findings might be used to help solve very real geographical dilemmas.

Transferable skills

Your geography degree is a journey through a field of study – learning about the history of the subject and its contemporary relevance. As part of your studies you develop specialist knowledge about people, place and space. This ranges from theoretical knowledge (how geographers have understood the concepts of space and scale over time, for example); empirical knowledge (relating to specific events, places and case studies); and practical knowledge (skills in GIS, interviewing, quantitative analysis, questionnaire design, and so on). However, another feature of your degree is the transferable skills that can be gained. These are skills that are not subject specific but can be applied outside of the degree. The dissertation is overflowing with transferable skills that will be useful for your future. These include (in no particular order): identification of a problem or gap in knowledge; synthesis of existing ideas; data management; ability to analyse materials and extract important information, patterns or trends; report writing; time management and planning skills.

Future employment

The dissertation allows you the opportunity to develop important skills that employers value in the workplace. It might be that the subject of your dissertation links to a particular job (say, if you were to complete a dissertation in an element of urban planning or environmental protection). More so, some dissertation topics may actually lead us to work with particular agencies, industries and businesses and these can be useful for gaining work experience, internships or future employment. But whatever the topic, it is also a piece of work that helps you develop skills from which you can draw in the future to demonstrate your competencies. When you apply for jobs, you can use your dissertation (and the experiences gained in researching it and writing it) to bolster your application and illustrate your skills.

Starting out: Top tips

We now know what a dissertation is and why we complete one, but how do we get started on the task? Before embarking on this book, I asked some of my own students what they wished they had known at the start of the process. Here's what they said:

'I wish I'd been more organised'

The dissertation is a piece of work which you complete over an extended period of time. It is easy to look at the deadline (often some months away) and feel there is plenty of time. But that time needs to be planned carefully so that it does not slip away. Although a dissertation should be completed bit by bit, it is also useful to try and envisage the whole process and what milestones you would hope to achieve at specific points of the journey. Typically there is a period at the beginning of a project that is spent designing your research. This includes reading around your topic; working out what data you need and figuring out how to collect it. This is followed by a period of time 'doing' the research, and thereafter a time which should be mapped out for analysis, writing up and delivering your project. Even if it is rough to start with, having a general picture of what you'd like to complete, and by when, can help you feel in control of your project. The timetable shouldn't be static though. You can hone and develop it as your project progresses. It is worth discussing your timetable with your supervisor at the outset to ensure it is realistic.

Go online! Visit **https://study.sagepub.com/yourhumangeography** for a blank copy of a timetable. You can download this and fill it in to plan out your time, or print it off to complete if you prefer.

'It was all sort of jumbled – I couldn't do the interview when I'd planned'

Whilst planning your time is a must, and whilst dissertation research generally follows a line of designing–doing–delivering (as the title of this book suggests), it is also good to know that research can be messy. Research often involves elements that cannot be foreseen or prepared for. When we research, we go into the unknown. The whole point of research is to study something that we know little about, or want to know more about. If we knew the answer, it wouldn't be worth doing. Therefore, it is useful to know that things change – our ideas, our abilities to collect data, the interpretations we bring to findings. Whilst we can attempt to impose order on things, research is often organic. As my student noted, in spite of planning for the interview, they couldn't conduct it when they hoped (instead they met the participant when they had timetabled to start writing up). Research then, requires us to be simultaneously organised, but able to adjust to change.

'I wasn't prepared when I saw my supervisor. I forgot what they said and then I felt I couldn't go back'

The dissertation presents the opportunity for engagement with academic staff in a way you probably won't have encountered prior to starting your project. A key feature of the dissertation process is one-on-one supervision (check your own institutional guidelines, as each department or school will differ slightly in how this is arranged). You will normally be allocated a tutor who will advise on your research. On occasion you may switch supervisors during the process due to staff changes or unforeseen circumstances. Your supervisor is the most important person for offering guidance. However, there are some general things that are useful to know about the supervisory relationship. Firstly, it is vital to remember that your project is an independent piece of work. Supervisors are there to supervise – or oversee – the project. Your supervisor can act as a soundboard to your ideas, offer reassurance (and critique), but they will not to do the project for you. Recognising this at the start is important for knowing what you can expect. Second, every student–supervisor dynamic is different. It can sometimes be easy to compare your experience with that of your peers but this can be unhelpful. All academic staff have differing methods of supervising that they have tried and tested over time. In short, there is no standard supervision experience that can be charted here on paper. It is important to put trust in your supervisor. Remember that they have ample experience in helping students design, do and deliver dissertations. Third, what you get out of supervision depends on what you put in. Always go to supervisory meetings prepared (for example, with a list of questions or issues you'd like to discuss). Ensure you have a means of taking notes (a paper and pen, or an electronic pad). Supervisors will discuss ideas with you, suggest readings, or will offer constructive criticism. In the meeting it can be hard to remember everything, so be sure to take notes. You will likely need to refer back to them.

'The marking criteria! I should have known more clearly how I was being assessed and what mattered'

In this chapter we have discussed the benefits of alleviating unknown entities, arming yourself with information, and being prepared. It is also worth taking a look at your department or school's marking criteria for the dissertation (see Task Box 1.1). Marking criteria are designed to inform you how any given assessment is marked. What are staff looking for when they read the project? By knowing this at the beginning you can ensure your dissertation covers all the bases that are essential to a successful dissertation (see also Information Box 12.2, p. 210 below). You can even use the criteria as a checklist as you write, marking your own work as you go. For example, do you think you have achieved what is required of a 2:1, or a First as specified on the mark sheet? Criteria look different at every university, so it's vital to find out how the project is assessed where you study.

'My tutor told me to do the project bit by bit. That really helped as I felt so overwhelmed by it all, by writing all those words'

On the one hand, it can be helpful to visualise the dissertation as a whole for the purposes of managing your time (as noted earlier). However, in order to reach that whole, the dissertation should be broken down (as the timetable also suggests). Doing a project bit by bit – working section by section, task by task, month by month – makes such a big piece of work manageable. When completing your timetable, break your project down into chunks that you can feasibly complete during a portion of allocated time (time for reading, time for researching, time for collation of data, analysis of data, and so on). It can also be helpful to produce more detailed timetables for certain parts of the project. For example, you may wish to complete a new timetable that breaks the writing stage up into manageable portions: allocating time to write each chapter; to draft and re-draft, and to format and print the work (see Chapter 11, 'Writing up').

This section is by no means an exhaustive list of top tips for the beginning of the dissertation process. You will no doubt be given more advice and different guidance when speaking to your supervisor and to students who have been through the dissertation process. However, one last piece of student insight should not be forgotten: "really, I just wish I'd enjoyed it more!" The dissertation is an opportunity to study something that you are passionate about and want to explore. Thus, try not to worry about it, but throw yourself into your work, and into the world.

How to use this book

This book follows the ethos set out above: that the dissertation is easier to complete if we break it down into parts. Whilst research can be a messy and complex process, we can split that process into three key stages – **designing, doing** and **delivering**. These constitute the three sections of the book to follow. However, this book can be read in varying ways. Certainly, you can read it cover to cover, in a linear fashion – using the book as a reference guide as you work through the journey of completing your dissertation – start to finish. Alternatively (and as well as) you can also dip in and dip out at different stages. Therefore, it doesn't matter if you are reading this and have already started your human geography dissertation. You can simply refer to the section that is relevant to wherever you are in the process. The contents page, alongside maps located at the start of each chapter, can be used to identify relevant sections. The book also includes a number of tasks to help frame your thinking, alongside examples from students who have recently completed dissertations (you will see these in the Graduate Guidance boxes). Also included is a host of links to online material that might be relevant to your project, indicated by the 'Go online!' icon, which you will see throughout. You can access this material by visiting the Companion Website. Here you will find links to selected Sage journal articles, alongside video clips, extra resources for research, copies of some of the key tasks that feature in the book and a checklist of 'take home' messages in relation to each stage of designing, doing and delivering a dissertation. In short, the book can be used in relation to *your* human geography dissertation. The book covers the following ground:

Section I: Designing

This section covers the task of formulating your project. We begin in Chapter 2, 'Starting out: Identifying your approach' by exploring how geographical concepts and wider theoretical ideas can provide a vital underpinning to your dissertation project. In Chapter 3 'Getting going: Finding a topic', we start to home in on the design process by examining the varying strategies that you can employ for finding inspiration for your dissertation topic. Chapter 4 'The next step: Developing your research question' considers the use of literature to identify a project that emerges from, and contributes to, an existing pool of knowledge. It also covers how you can narrow down broad ideas into specific, workable projects. The chapter also explores how we turn our topic idea into a usable set of research questions, aims and objectives. In Chapter 5, 'Final preparations: Is your project workable?', we turn to issues of practicality and feasibility. Once you have a research idea you must ensure the project is viable. This chapter covers the business of ethics, health and safety, and legality. How can you ensure your research is responsible, sensitive and safe? This chapter considers how you take on the role of a professional geographer, ensuring your research meets the standards expected when engaging with the relations between people and place.

Section II: Doing

This next section of the book covers the process of completing the research you have designed. Chapter 6, 'Doing reflexive research: Situating your dissertation', begins by considering how we do research reflexively. In human geography there is now an appreciation that who we are shapes what we do and how we engage with and make sense of the world. When we conduct our research (as well as when we plan it) we must take into account our positionality and how this informs the knowledge we produce. This chapter considers how we account for the factors that impact our work and the benefits that this process of accounting can bring to our data collection and analysis. Chapters 7 and 8 deviate from more traditional chapter formats to instead offer an overview of the key methods employed by human geographers. Chapter 7, 'Making research happen: The methods glossary', outlines a suite of conventional methods used in geographical research (from interviews, focus groups, ethnography and textual analysis, to statistical analysis and GIS). Chapter 8, 'More on methods: Approaching complex social worlds', continues in this vein, highlighting contemporary methods, including participatory and action-orientated approaches; creative and experimental modes of researching; sensory and mobile methods, alongside approaches to researching online. These chapters feature detailed lists of further reading relating to each method which should be consulted in critically understanding and engaging with the use of varying techniques in geographical research. Chapter 9, 'Selecting your methods: How to make the right choices', completes the section by investigating how we choose appropriate methods in respect of the research questions asked. What methods are suitable and justified for your project? Which methods are not, and why? This chapter helps us move away from simply selecting methods we like, to those that actually enable us to gather the information and materials we require to respond to our aims and objectives.

TASK 1.2

The research diary is a longstanding tool of the geographer. It provides a way of recording observations about the world, but also a means of reflecting on personal engagements with the research process. In recent years, scholarship that attends to student learning in geography and related disciplines has identified the benefits of using diaries for self-reflection (see Dummer et al., 2008).

As part of the dissertation process, keep a diary of your progress. Your diary might be a paper file, folder or notepad or it could be an electronic document on your tablet or computer. Use your diary for two purposes. First, keep a record of everyday things related to your work, for example ideas for projects; suggested readings; notes from supervision meetings; key words and phrases, and so on (see Task 2.1 in Chapter 2). Second, use your diary to keep a record of your experiences, encounters, thoughts, feelings, emotions; a note of things that went well, or went badly; moments that surprised you, or confirmed your thinking during the research process. Diaries can act not only as a means of jotting down important points to remember, but can also be another source of data, a reflexive account of the work you conducted which you can draw upon in relation to your research question. Amy Jones explains the benefits of keeping a diary in the Graduate Guidance box below.

KEEPING A DIARY ON THE WALES COAST PATH
AMY JONES

My research focused on the Wales Coast Path, an 870 mile continuous path that follows the whole of the Welsh coastline. I sought to explore the embodied act of walking and what it meant to be able to walk the coast of an entire country. My fieldwork consisted of joining people on their walks and in total I took part in 40 walks, covering over 400 miles (and I had the blisters to prove it!). As well as interviewing people whilst walking, I kept an in-depth field diary which I wrote up the evening after each walk, based on short note-taking throughout the day.

But why did I keep a diary? First it was a tool of research in its own right. It allowed me to capture the embodied experience of walking that was central to my research question. It also supplemented my interviews by adding extra layers of detail to my explorations. For example, the walkers I interviewed stressed the freedom that walking offered them compared to the restrictions of everyday life. Yet the Wales Coast Path seemed to also have its own constraints. Walkers were committed to walking *every* step of the path. Observations recorded in the diary showed that walkers felt bound to the path and were unwilling to leave it to explore what was beyond:

> The path took us inland through woodlands where I knew of a statue which chronicled Carmarthen's maritime history. As the path meandered around the location of the statue I suggested we stop to see it. Yet the walkers had no interest in veering away from the path and merely pointed at the direction the Wales Coast Path sign guided us and continued purposefully along the path. (Walk 27: Llansteffan to Carmarthen, 9 miles, 18/10/2014)

The diary also allowed reflection on my positionality as a walker. It chronicled how my identity as a walker changed throughout the fieldwork. For example, at the beginning of my research I was clearly seen as the outsider in the group.

(Continued)

(Continued)

At this time I hadn't walked much of the Wales Coast Path. I wasn't included in the pre-walk camaraderie and was often left holding the cameras for photographs. The field diary narrative showed that it wasn't until I had proven myself as a walker – and that I could keep up – that the walkers became unguarded in their conversation. The diary also allowed me to reflect back on how time changed my walking experiences. As I noted a year after my first walk:

> ...with hundreds of walked miles under my belt I'm no longer seen as simply a student but a bonafide member of the exclusive Wales Coast Path club. [When] [t]he time came to take the obligatory group photograph, the cameras were no longer handed to me but to the newest addition of the group who had yet to prove his walking prowess. (Walk 42: Amroth to Tenby, 7 miles, 02/05/2015)

Keeping a research diary was an intrinsic part of my project. Recording observations from the very beginning of the fieldwork provided important and rich research data. It highlighted unexpected findings and demonstrated that sometimes there can be a disparity between what people say and what they do. It also chronicled my own development as a researcher and gave me a fascinating insight into how I changed throughout the fieldwork process. Therefore, consider the benefits of using a diary in your own research. It will likely enrich your data with intriguing findings, and like my diary, it may reveal things you didn't expect.

Go online! Visit **https://study.sagepub.com/yourhumangeography** to access Sage journal articles related to the practice of keeping a research diary for your own human geography dissertation.

Section III: Delivering

The final section of the book covers how we deliver the dissertation in its final form. The section begins with Chapter 10 'Dealing with data: Approaching analysis'. This chapter covers how we forge a bridge between data and knowledge, interpreting our findings in order to respond to the research questions we have set. Chapter 11, 'Writing up: Where to start and how to finish', moves to the task of translating our research to paper. It covers what to write, when to write, how to write and where to write. Chapter 12, 'The last hurdle: Final considerations', explores how you reach the finish line of your dissertation. This chapter details the additional pages you must add to your written document; ways to proof and polish your work, as well as demystifying the marking process. This chapter also considers life beyond the dissertation and how it can help you in regard to future careers.

N.B. How not to use this book

It might seem strange to close the chapter by noting how not to use this book. However, this book does come with some caveats. This book does not (and could not) provide a 'holy grail' of dissertation advice. It does provide general guidance on good practice and covers common concerns and issues faced by human geographers. However, given the unique, individual and independent shape of each research project, it cannot answer every query. The most important source of advice on your project will always be your supervisor and their tailored advice in relation to your project. Moreover, whilst this book will tell you what a dissertation is, why we would do one, what skills we might require to succeed, and how we might complete the task – crucially, it can't (and won't) provide you with suggested topics or questions for study. As Bennett et al. aptly note,

> The topics … geographers can choose from are virtually infinite. That is why [it's best to] not generate a list from which a selection could be made. It would be meaningless, and might go something like: 'young male street fashion … depiction of urban malaise in comics … dog ownership and the use of open spaces … postmodern architecture in shopping malls … West Indian poets in London … the arts in place promotion …women as buyers and sellers in antiques markets. …' The two big questions in topic selection are 'Why do I want to know?' and 'Is it possible for me to find out? (Bennett et al., 2002: 93)

As the title of this book suggests, it is *your* human geography dissertation. And whilst you can use resources – this book, your supervisor, your peers, and so on – to support the process, the most exciting element of the dissertation is that ultimately it is your journey to complete. And, as previously noted, this is what it means to become a geographer – to make a contribution to our understandings of places and spaces around us, and to go into the world and explore it, question it and maybe even to change it.

Chapter summary

- Research is a process of identifying a problem or issue that is unanswered or requires greater attention. It emerges from, and contributes to, existing bodies of work. Research is the process that *makes* knowledge. When you conduct a dissertation – your own piece of research – you transition from being a student of geography to a producer of geography.
- A dissertation is an extended piece of work that you complete independently. It should include data that you have collected to answer the research questions you have set. A dissertation should make a contribution to a specific topic area, as well as enabling you to gain a host of transferable skills.

- Although we have to complete a dissertation as a compulsory element of our degree, it provides the chance to follow our curiosity, to be inquisitive and to delve into a subject, topic, issue or debate of real interest to us. Dissertations allow us to gain a sense of personal achievement and can even help us gain future employment.
- As this book suggests throughout, it is best to approach a dissertation by being prepared. At the start, arm yourself with information about the process, and, crucially, be organised and use a timetable. As you progress, tackle the dissertation bit by bit through the three stages of designing, doing and delivering.

Key readings

Burkill, S. and Burley, J. (1996) 'Getting started on a geography dissertation', *Journal of Geography in Higher Education*, 20 (3): 431–7.

Clifford, N., French, S. and Valentine, G. (2010) 'Getting started in geographical research', in N. Clifford, S. French and G. Valentine (eds) *Key Methods in Geography* (second edition). London: Sage. pp. 3–15.

Cloke, P., Cook, I., Crang, P., Goodwin, M., Painter, J. and Philo, C. (2004) 'Changing practices in human geography: An introduction', in P. Cloke, I. Cook, P. Crang, M. Goodwin, J. Painter and C. Philo (eds) *Practising Human Geography*. London: Sage. pp. 1–34.

SECTION I

DESIGNING YOUR HUMAN GEOGRAPHY DISSERTATION

2

STARTING OUT: IDENTIFYING YOUR APPROACH

CHAPTER MAP

- The connections between thought and practice
- Finding your approach to geography
- Theoretical approaches for geographers
- The geography in your dissertation
- Philosophical ruminations: Moving forwards

The connections between thought and practice

As a student of human geography you are likely beginning to realise that the discipline, as well as having a 'real-world' relevance (with application to society, culture, politics and the environment), also has varying philosophical and theoretical underpinnings that shape our approaches to examining such areas of study. You may have attended lectures as part of your degree that seem more appropriate to the study of philosophy rather than geography. So what is the connection between geography and the wide array of concepts and theories that we are taught as part and parcel of a geography degree? It is argued that the process of conducting or doing research can never truly be separated from an understanding of and engagement with philosophy (or as Aitken and Valentine more intelligibly put it (2006: 1) 'ways of knowing'). An example illustrates the point.

Two students are embarking on dissertation work. Both are interested in the role of the internet and virtual technologies for socio-spatial life (see Kinsley, 2014 for inspiration). However, both are interested in asking very different questions about the subject matter. One knows she wants to explore how internet usage varies spatially – particularly between urban and rural areas, and in relation to a variety of demographic characteristics (such as age and income). The other student knows he wants to investigate how internet usage impacts socio-spatial experience. He wants to ask how we engage with virtual space, the ways it compares to 'real' space, and what opportunities it affords for users (see the Graduate Guidance box below).

These students have each selected – perhaps without realising it at the time – two alternative ways of knowing the world. The first takes what might be called a **realist** approach (see Couper, 2014: 227). For realists there is an understanding that the world already exists out there for us to examine, to hypothesise about, to map, to count and to do tests on. The second student takes an **anti-realist** approach. For anti-realists there is an understanding that the world is not just out there to be examined, rather it is only knowable and understandable because we invest meaning in it. One ascribes to **objective** knowledge, which provides verifiable results in the form of statistics and maps. The other subscribes to **subjective** knowledge, which provides specific and partial results in the form of data that unlocks experiences, beliefs, desires, values (see Information Box 2.1, p. 23 below). The students are also shaped by different theoretical positions (one by positivist and post-positivist approaches, the other by debates in post-modern thinking; see the section on theoretical approaches to follow). So – whether they are conscious of it or not – both students are thinking about philosophy and theory in *doing* their research. As Graham (2005: 11) puts it, 'even the most philosophically inarticulate researcher makes philosophical choices simply by doing research'.

The good news is that you will often unwittingly veer towards a particular 'way of knowing' the world without even realising it. As Shurmer-Smith notes (2002: 11), the perspectives we can engage easily with, and ideas we subscribe to, will shape the way we look at and 'know' spaces and places around us. But why does all this matter? Graham argues that we should make an active effort to 'bother' about and uncover the links between thought and practice as this will help us to forge stronger research projects (Graham, 2005: 11). All good dissertations have two connected components – a strong empirical focus (*what* you study) and a solid, appropriate philosophical grounding (*how* you study it). Appreciating how you think about the world will ensure that you have situated your project in the right literature and debates (Chapter 4) and have used the right methods to examine your question (Chapter 9).

But how do we actually begin to think seriously about 'ways of knowing' in regard to your human geography dissertation? In this chapter we begin by considering how you find your own approach to geography. We do so by exploring the two key 'ways of knowing' that shape work in the discipline: knowing that is objective (seeking certainty and fact) and knowing that is subjective (seeking individual and personal realities). Next, the chapter homes in on more specific ways of knowing, examining some of the core theories that have shaped human geographic thought. These are the 'isms' that often underscore our approaches – positivism, humanism, Marxism, feminism, post-modernism and post-humanism. Finally, the chapter highlights the importance of ensuring your approach is geographical. Here we explore how to connect your project with some of the fundamental concepts that underpin the discipline: space, time and movement. The chapter closes by ruminating on the importance of identifying your approach, before moving on to the task of designing your dissertation in Chapter 3.

Finding your approach to geography

Geography is often said to be an eclectic discipline. It does not subscribe to a stable, structured sense of identity. As Richard Peet has noted (1998: 1), 'geography has a permanent identity crisis because what geographers do is complex'. The good news is that this makes the discipline 'dynamic, interesting and intellectually fertile' (Peet, 1998: 1). Geography encapsulates everything from 'hard' science (the testing of hypotheses and modelling of phenomena), to the arts and humanities (subjective and qualitative studies of human engagements with the world). Accordingly, geographers (especially human geographers) are a motley bunch, exploring space, place and time in a variety of quite differing ways.

When embarking on a dissertation, you first have to make a decision on which kind of geography you hope to explore (working on the assumption there is not one, single, geography). Your choice as to which brand of human geography your dissertation falls into (cultural geography, political geography, population geography, environmental geography) is – inadvertently – a philosophical choice. It is not simply a preference between different types of human geography, but is underscored by how we look at the world around us. These different types of geography are situated in differing philosophical and theoretical traditions, so being alert to this is crucial. In turn, you can then plan research, use methods and employ analytic techniques that are in-keeping with the area you are studying. It is useful to consider the vast array of human geographies that are undertaken in your own department as a starting point for identifying your own approach to study (Task 2.1).

TASK 2.1

Look at the Geography home page of the university where you study. Explore the human geography research that is conducted in your own department or school. You will usually find this under the tabs describing research (most departments and schools have research groups and clusters) and on individual staff pages. Make rough notes on the following:

- What different types of geography (cultural, political, environmental, etc.) are studied?
- What particular topics, spaces, places, times, objects or subjects are under examination?
- What different techniques and methods are used?
- What kinds of questions are staff asking? How are they asking them?

(Continued)

(Continued)

Once you have jotted down the areas of research interest, look back on your notes. Begin to draw lines of connection based on the following questions:

- Which topics and areas of study seem to link together?
- Which geographers are asking the same kinds of questions? Is there a pattern?
- Which geographers are using similar methods for their research?

Once you have grouped things together, ask yourself the following:

- Why are some types of human geography different from others? What makes them so?
- Why do geographers ask the questions they do about their topics? What are they hoping to find out?
- Why do you think the methods used are the same or different?

Go online! Visit **https://study.sagepub.com/yourhumangeography** for a link to the University of Liverpool Geography home pages. Notice the range of research that academics conduct in this department. Some staff fall under the remit of population geography (exploring socio-spatial inequalities, housing segregation, health and wellbeing indicators and the correlation between place and crime). Others fall under the banner of social, cultural and political geography (exploring the experiences of Polish migrants in the UK; the use of alternative currencies to foster community activism; and the role of the body in research practice). How are these kinds of research underscored by different ways of knowing the world?

The point of the task is to illustrate that the work of human geographers can be very varied. In my own department, and those I have worked in previously, the research conducted by human geography staff is incredibly diverse. Some of my colleagues have been engaged in research that seeks to visualise and map community and population changes over time using Geographical Information Systems (GIS). Others have drawn on Census data collected by the government about the UK population (along with other regional and national surveys) to make statistical correlations about location and a host of socio-economic factors (educational attainment, experience of poverty, mortality rates, health, job success, and so on). Most recently I have worked with colleagues who have sought to explore how governments and non-governmental organisations respond to disasters, drawing on official documents, policies and manifestos. I have also worked alongside colleagues who have used archive records to piece together stories of our colonial past in the present. My own work has employed interviews to speak first hand to people about their experiences running a pirate radio station (Peters, 2011a).

Some of these topics and areas of research have things in common, others are notably different. A geographer conducting a statistical analysis of electoral data for wards in England and Wales to understand voting behaviour for example, would appear very different from a geographer using interviews and focus groups to do the same job. But why do we know these modes of research are somehow distinct from one another? It comes down to our approach to geography and how we view the world. In other words, do we see:

1. A world that exists out there already, which can be measured, recorded, quantified, to reach objective, factual, verifiable knowledge claims.
2. A world that we are part and parcel of making, which is messy and complex, from which we can make specific, situated, partial claims to knowledge.

In other words, when we approach research we need to ask whether we want to reach statistically verifiable claims about geographical phenomena, or if we want to understand the diverse meanings that permeate human relations with space and place (or if our way of knowing is open to both; see Sui and DeLyser, 2012, and Chapter 9, 'Selecting your methods'). Crucially, when starting out with your dissertation, it is worthwhile asking if you want to arrive at objective or subjective knowledge (Information Box 2.1).

INFORMATION BOX 2.1
OBJECTIVE AND SUBJECTIVE KNOWLEDGE

Traditionally, **objective knowledge** is that which is unbiased and impartial – it appears to be fair and non-partisan. As such, objective knowledge seems 'god like' – it seems to have come from a higher power. Donna Haraway's pivotal work (1988: 581–2) notes how objective knowledge is 'distanced' from those who create it (researchers) and, as such, it has an indisputable quality. Objective knowledge has validity in that it is based on 'facts'. It therefore makes grand claims about what exists. Objective knowledge is often – but not always – linked to forms of **positivism** (which can be traced back to the work of Auguste Comte). Positivism refers to a theory that states knowledge can only be verified and guaranteed as 'certain' through rigorous empirical testing 'to reach positive knowledge' (knowledge that cannot be challenged – unless through similar positivist testing) (Peet, 1998: 23). Moreover, objective knowledge tends to be nomothetic, that is, generalisable and universal. Objective knowledge (or the means through which objective knowledge emerges – hypotheses, formulas, models, and so on) can often be applied to different examples. Often the work of spatial scientists (creating mathematical models in relation to space) is said to be objective. That said, many geographers who work with statistics, maps and models today, although still seeking to make objective knowledge claims, take a **post-positivist** approach whereby it is acknowledged that certain biases shape results and findings.

(Continued)

(Continued)

Subjective knowledge takes into account the subject – the person – the intricacy and complexity of their thoughts, feelings, and emotions. Here it is accepted that there cannot be one single, objective way of knowing the world. Instead there might be many ways of knowing the world. Subjective knowledge came to the fore with approaches such as humanism, which argued that human experience of the world cannot be understood with (or contained by) models, hypotheses and formulas. Humanistic geography sought to re-people the discipline in the face of spatial scientific approaches, where the intricacy of human life seemed all but absent. This required a different form of knowledge, which reflected the variety of **experience**. Unlike objective knowledge, subjective forms of knowledge are multiple (and do not seek one answer, or to make generalising claims to understanding). Subjectivity allows us to acknowledge there might be multiple ways of knowing. In this sense, subjective knowledge is idiographic – that is, it is unique, personal, and specific.

Have a think now. What type of human geography sits well with your own sensibility? Do you relish numbers and the ability to map phenomena to space, or to correlate human characteristics (age, health, ethnicity) to particular places, at particular times? On the other hand, are you excited by the diversity of meanings that can be unpacked in trying to understand some sort of socio-spatial experience or phenomenon? Identifying your own approach (your own way of knowing the world) will help in honing the kinds of questions you want to ask in a dissertation, and, in turn, the sorts of methods you will use. So when setting out with a project, it is essential to consider how you think about the world. This is because thought is connected to practice.

Go online! Visit **https://study.sagepub.com/yourhumangeography** if you would like to read more about philosophy and theory. Here you can find additional material outlining how 'ways of knowing' underscore ways of doing.

Theoretical approaches for geographers

Whilst there are overarching 'ways of knowing' that support how we approach research, there are other considerations you must make when starting out with an undergraduate dissertation. These relate to the influence of theory. Geography has been shaped by theories that have guided the social sciences more generally. Notably these theories have been adopted within the discipline to develop distinctly spatial innovations relating to such ideas (for example, post-colonial theory has been utilised by geographers who have consequentially demonstrated how imperial dominance has had unique spatial outcomes – see Davies, 2013; Jazeel, 2014;

Sidaway, 2000). Just like ways of knowing more generally, often student projects are located within specific theoretical approaches without us realising. As Tim Cresswell describes,

> [i]f we choose to look at the micro spaces of the home, there is a history of feminist theory urging geographers to take private space seriously. If we choose to study the structuring of public space, there are any number of theorists who have argued about the meaning of 'public' (let alone the meaning of 'space'). (Cresswell, 2013: 4)

Good dissertation projects should show an awareness of the theoretical stance that has shaped the research. It is worth asking if there is a specific theoretical tradition in which your work sits, which will help you better understand and situate your findings. This section provides – as a starting point – a simple overview of the key theories that have shaped geographical research, past and present. Reading these will provide the basis for locating your project within an appropriate theoretical context. However, you should also consult more detailed texts and there is now a host of recent books dedicated to the task of exploring geographic theory (key publications are listed below and articles that use each approach are available on the Companion Website). The sections to follow also list some 'classic' texts associated with each theoretical shift, should you want to develop your reading further. The theories charted here follow a broadly chronological order in respect to their adoption in geographical thinking. That said, one theory rarely replaces another – and traces of previous frames of thinking always prevail.

Key readings

Aitken, S. and Valentine, G. (eds) (2006) *Approaches to Human Geography*. London: Sage.

Couper, P. (2014) *A Student's Introduction to Geographic Thought: Theories, Philosophies, Methodologies*. London: Sage.

Cresswell, T. (2013) *Geographic Thought: A Critical Introduction*. London: Wiley-Blackwell.

Nayak, A. and Jeffery, A. (2011) *Geographical Thought: An Introduction to Ideas in Human Geography*. London: Pearson.

Go online! Visit **https://study.sagepub.com/yourhumangeography** to access a series of Sage journal articles that show how each theoretical approach (Positivism, Humanism, Marxism, Feminism, Post-Modernism and Post-Humanism) have been employed by human geographers to help them understand their topic area.

Positivism

Positivism has a long tradition in geography. This includes innovations such as Alfred Weber's 'Industrial Location Theory' (1909), which mapped the prime locations for industry based on minimising costs of production and distribution (Johnston, 1986: 37) and Walther Christaller's 'Central Place Theory' (1933) that explored 'the size, function and distribution of settlements' – making broad statements about population and the provision of services (Johnston, 1986: 35). That said, in the 1960s the discipline turned firmly to positivist approaches, joining a quantitative revolution which saw geographers employing models and formulas to develop laws and equations that could compare places and predict spatial outcomes. In the mid-twentieth century, Regional Geography, which preceded this broader positivist shift, saw the discipline engage with a descriptive approach of classifying regions, developing a 'comprehensive synthesis of everything in a given area' (Couper, 2014: 18). Such an approach could provide a detailed study of specific regions but it could not say how geography was more widely applicable. Accordingly, geographers sought to make their discipline 'more scientific, nomothetic (law-stating) instead of idiographic' (Peet, 1998: 22). They developed positivist approaches based on empirical (that is, direct) observation of the world (also see Information Box 2.1, p. 23). A positivist approach arrives at 'truths' because the findings are said to be objective – mathematically certified and sound – and based on data drawn directly from the world. Positivism shaped an era of geography dedicated to **spatial science**. Although geographers still use quantitative approaches to arrive at objective knowledge (see Information Box 2.1, p. 23) many now position themselves theoretically as 'post-positivist' in their approach. Whereas positivist approaches claim absolute certainty in findings, **post-positivist** research appreciates the factors that shape any knowledge claim (for example, data limitations), thus giving the findings produced great legitimacy (see, for example Kwan and Ding, 2008).

Key readings

Chorley, R. and Haggett, P. (eds) (1967) *Models in Geography*. London: Methuen.
Harvey, D. (1969) *Explanation in Geography*. London: Edward Arnold.

Humanism

Humanism emerged in the early 1970s as a direct response to the quantitative and positivist leanings of geography at the time. Geographers adopted it as an approach that focused on 'the individual as a thinking being' (Johnston, 1986: 55). A geography inspired by humanistic principles worked to re-people geography and make the discipline more **idiographic** (focused on the specific and personal) to counter approaches that focused on the generalisable and universal. As Holloway and Hubbard note, for humanists, there is no world outside of human experience;

we make the world what it is through our experience of it (2001: 71). In other words, general models and laws are totally suspect as they don't reflect the way we actually live our lives (Holloway and Hubbard, 2001: 71). Humanistic geographers therefore seek to understand the intricacies of human **experience** – what it is like 'being in the world'. To do so they draw greatly on **phenomenology** as a theoretical approach. Phenomenology argues that phenomena in the world only come into existence through human experience of them. Experience is key to humanist geographers because it is via individual experience and the meanings it generates, that place is forged and formed. Subsequently, the concept of **place** (as space made meaningful via human engagement) came to the fore (Tuan, 1977; Cresswell, 2004). However, whilst humanism was a persuasive alternative to the mathematic models and generalising laws that characterised spatial science, it was also critiqued for a lack of politics and a failure to account for the differentials of experience 'in place' that were forged through processes of power. The home, a core site of focus for humanists was regarded unproblematically. For geographers like Tuan (1977) the home was a site of attachment, coziness and comfort. The work of bell hooks (1990), the African-American cultural theorist, along with feminist proponents (see Rose, 1993) has contested the comfortable spatial associations made by humanistic geographers. Whilst humanistic geography is now less fashionable, its interests remain as geographers continue to explore human experience and engagement through approaches typically branded **post-phenomenology** (see Ash and Simpson, 2016). Here geographers are interested in opening up a more **critical** engagement with embodied, lived experience (see Edensor, 2000; Simpson, 2011; Wylie, 2005).

Key readings

Buttimer, A. (1976) 'Grasping the dynamism of the lifeworld', *Annals of the Association of American Geographers* 66 (2): 277–92.

Tuan, Y.F. (1977) *Space and Place: The Perspective of Experience*. Minneapolis: University of Minnesota Press.

Marxism

Theories derived from the work of Karl Marx (and his colleague Friedrich Engels) gained momentum in geography in the 1960s, forging what would become known as a **radical** era of geography (for a detailed introduction see Peet, 1998: 67–146). This period marked the beginnings of a socially engaged and proactive geography. Where Marxist ideas had no particular spatial leanings, geographers (such as David Harvey, Doreen Massey and Neil Smith amongst others) recognised that it could provide a new framework for understanding relations between people, space and place. The employment of Marxist thought has been varied (there is not one Marxism, but differing forms – classical Marxism, structural Marxism, feminist Marxism and now post-Marxism and so on). However, broadly, geographers have worked with two central ideas from Marx's writings. The first relates to the emergence of particular forms of **social organisation**. Marx noted that in capitalist societies the key division is between working and ruling classes. The working class (the proletariat) have no economic independence and must work for rulers (the bourgeoisie), those who own the means of production (the factories and equipment). Inequality lies between those who own the means of production (the machinery and buildings of a factory, for instance) and those who have nothing more to sell than their labour. In this system there is a mutual dependence – the working class have to work, the ruling class need them too, so an exploitative cycle ensues. Geographers have used Marx to argue that class conflict and uneven development are perpetuated in space. For example, David Harvey (1973) has argued that capitalist dynamics shape space: new centres come into being whilst others fall into decay. Uneven development occurs when capital is shifted and reinvested around the globe. Capitalist processes exploit new spaces to maintain profitability, leading to growth in some areas with other areas left 'behind' through capitalist exploitation. Secondly, and more recently, geographers have used Marx's ideas relating to **commodity fetishism**. In *Capital*, Marx (1867: n.p.) has written that commodities are the products of nature. However, when nature is transformed into commodities through processes of production, 'it is changed into something transcendent' with a 'mystical character', disconnecting the product from its origins. Consequently, the origins of commodities – the fact they are things of the earth – becomes masked. We cannot see things for what they really are. Geographers have drawn upon, and critiqued this work, to 'unveil the fetish' and trace the complex geographical networks relating to material items; 'following the thing' (see Cook, 2004).

Key readings

Harvey, D. (1973) *Social Justice and the City*. London: Edward Arnold.
Massey, D. (1984) *Spatial Divisions of Labour: Social Structures and the Geography of Production*. Basingstoke and London: Macmillan.

Feminism

Feminism, Tim Cresswell has recently noted (2013: 147), has appeared to many to be a 'dirty word'. Feminist ideas, it seems, take us to uncomfortable ground. Yet feminism is a 'powerful collection of ideas and practices' that cannot, and should not be ignored (Cresswell, 2013: 148). Generally, it is defined as

> a political movement that seeks to overturn gender inequalities between men and women. ... Feminism challenges and resists the gender roles and relations that position men and women in different and unequal ways in society. As such, feminism is concerned with ... how all spheres of life are gendered in particular ways. (Blunt and Wills, 2000: 90)

The basis of feminism is an attention to the ways in which **gendered relations** are unequal. Feminism has emerged in different 'waves' over time, the first relating to political movements for women's rights and the second to the more pervasive and ingrained forms of discrimination that shape everyday life (see Nayak and Jeffery, 2011: 130–2). However, like Marxist approaches, geographers have explored the ways in which unequal gender relations are perpetuated and reinforced through *space*. Geographers have been part of larger debates around the idea of gender as a social construct (see Butler, 1990) and explorations have investigated the spatial performances of gender (see McDowell, 1997) and representations of gender (Nash, 1996). For example, Gill Valentine's early work on fear in the city (1996) explored how the urban realm has been threatening to women, consequently shaping their spatial movements. Linda McDowell's seminal work on gender and the workplace (1997) investigated how women altered their appearance and behaviour in city companies, taking on more masculine characteristics in order to challenge discrimination in these unequally gendered landscapes (see also McDowell, 1999). Geographical work has also expanded to take the spatial representations and performances of masculinity seriously (see Hopkins and Noble, 2009) and to interrogate the gendered formation of geographical knowledge. This has been, in part, a questioning of the role of women in the discipline (Domosh, 1991). Moreover, following the broader critiques of objective knowledge (see Information Box 2.1, p. 23 above), geographers have argued for the situated nature of knowledge (the idea that all knowledge production is shaped by the creator) (Rose, 1997). This has led to the development of distinct feminist methodologies (Moss, 2002; see also Chapter 6).

Key readings

Domosh, M. (1991) 'Towards a feminist historiography of geography', *Transactions of the Institute of British Geographers* 16: 95–104.
McDowell, L. (1997) *Capital Culture: Gender at Work in the City.* Oxford: Blackwell.

Post-modernism

Over the past couple of decades, 'geography' has changed its name. Where once book titles and articles in academic journals spoke of the 'geography of crime' or the 'geography of climate change', scholars working in the discipline began talking about the subject in the plural rather than the singular. Geography became *geographies.* Take a look in a recent issue of any key geographical journal (for example *Progress in Human Geography*) and you will see that authors tend towards using the term 'geographies' when discussing spatial phenomena. This shift has its roots in post-modernist thinking. Post-modernism separates a given field of investigation from grand, organising structures. In other words, there is not a singular causal link between a phenomenon, event, place, person, and an overriding narrative that explains it. Rather post-modernism is concerned with a world that is multiple and complex. A host of philosophers writing at the time (such as Deleuze and Guattari (1988)) rejected, as Peet puts it (1998: 195) 'modern assumptions of coherence and causality entirely, arguing instead for fragmentation, multiplicity, and indeterminacy'. This shift has created a geography sensitive to difference (note other 'turns' (theoretical and methodological shifts) such as **post-colonial** thinking – see Nayak and Jeffery, 2011; see also Chapter 11) and attuned to how a diverse array of humans – young, old, rich, poor, black, white, straight, gay – experience space and engage with space differently (see Peet, 1998: 171). However, some geographers challenged this move towards a fragmented geography, arguing instead for a need to return to some sort of order (e.g. Dear, 1988).

Key readings

Dear, M. (1988) 'The post-modern challenge: Reconstructing human geography', *Transactions of the Institute of British Geographers* 13: 262–74.
Harvey, D. (1989) *The Condition of Postmodernity*. Oxford: Blackwell.

Post-humanism

Human geography is, most obviously, concerned with human relations with space and place. However, in recent years, the discipline has made a 're-turn' towards subjects that are **non-human** (materials, commodities, animals, and so on) or **more-than-human** (nature, climate, atmosphere, etc.). This 'post-human' turn has been driven by the acknowledgement that geography has become too anthropocentric (human-centred), omitting examination of other, non-, and more-than-human things and beings (see Panelli, 2010). As Sarah Whatmore has noted, there is a need for human geography to reconnect with life beyond the human (2006: 601). Geographers are questioning 'how humans and non-humans relate' and are employing new methods to explore 'the ways in which humans, animals, plants and other actors and intermediaries come together' (Bear and Eden, 2008: 488). These investigations are incredibly varied in scope. On the one hand, geographers have increasingly investigated **commodities**; the biographies of objects and the 'force' of material things (as Jane Bennett puts it, 2004). Here the path of an object can be followed to unlock the geographies enfolded within and spun from it (see Cook, 2004). Additionally, objects have been understood to hold a power, or enchantment, that has affects when in touch with human life (Bennett, 2004). On the other hand, geographers have centralised nature, breaking down the false dichotomy between nature and culture that has typified traditional examinations, instead speaking of a **social nature** (Castree and Braun, 2001). Here nature (in its many guises, from the countryside, to parks, to gardens) is held in *relation* with, or seen to be *hybridised* with humans (Whatmore, 2002).

Geographers have also investigated human relations with animal life (see Buller, 2013; Philo and Wilbert, 2000). More recently, geographers have continued this engagement with the non- and more-than-human world, to consider the **elemental geographies** of soil, air, water, fire (see Anderson and Wylie, 2009); and the ways in which these 'earthly' features are co-combined with human spatial experience.

Key readings

Jackson, P. (2000) 'Rematerializing social and cultural geography', *Social and Cultural Geography* 1 (1): 9–14.

Panelli, R. (2010) 'More-than-human social geographies: Posthuman and other possibilities', *Progress in Human Geography* 34 (1): 79–87.

TASK 2.2

Geography can be a confusing discipline. It has its own specialist terminology which draws on an existing range of complex theoretical ideas. In Chapter 1, Task 1.2 suggested that you keep a diary documenting the process of your research. This works both as a tool for ensuring a critical approach to research and in writing your methodology. In this diary, or in a separate notepad (or computer document), you should have a section where you note down any new terms, concepts and theories that you come across during your dissertation work. Look these up (either in a standard dictionary, or a specialised text such as *The Dictionary of Human Geography*) and jot down the associated meanings. This will help you to build a knowledge of a wide range of terms and concepts as you develop your dissertation ideas.

The geography in your dissertation

In this chapter we have recognised that thought and practice are entwined and that our approach to geography shapes the kind of questions we want to ask, the knowledge we want to uncover, and in turn the methods we will use (see Chapters 7, 8 and 9). We have also seen that a wide range of theories underpin geographical work (and our own project will likely be contextualised within broader theoretical frameworks). Finally, though, when starting out, we must also ensure that our project is geographical. This may seem like an unusual statement. You are doing a human geography dissertation after all. However, given the diversity of human geography (as previously mentioned) – spanning from positivist approaches to post-modern investigations – the geographical elements of a project can sometimes get lost in the processes of planning research. Indeed, geography has become a catch-all term to refer to any phenomenon that happens in space. We might be forgiven in thinking that geography is everywhere and will therefore be implicit in any project we complete. But we shouldn't just assume our project is geographical, because geography is supposedly 'everywhere'. Whilst geography may be inescapable, we still must show how geography matters to what we are researching. This means we have to engage with the core interests (as I have called them) that underscore the discipline. Listed here are three broad areas that are crucial to the work of geographers: space and place; time and temporality; and movement and mobility.

Space and Place

If the core focus for sociologists is society; and historians, time; it is perhaps little wonder that space and place should be the central concern of geographers. As Derwent Whittlesey, a geographer at Harvard during the early to mid-part of the twentieth century said: 'space' is the 'basic organizing concept of the geographer' (1954, as cited in Blaut, 1961: 1). What is essential in an undergraduate dissertation is that space, place and *spatiality* feature. But how do we ensure our dissertation is geographical in this sense? Space and place refer to the dimensions, planes and pockets of the world in which we live. **Spatiality** refers to that which *relates* to space and place. For example, an exploration of the spatiality of living arrangements in turn-of-the-century tenement blocks would investigate how people lived in relation to space and place (the tenements, estate, city, and so on) in a given time period.

But how we think about spatiality depends on how we understand the concepts of 'space' and 'place'. Traditional understandings have conceived of space through an **Euclidean** framing. This takes space to be something geometric – a measurable plane or container on and in which social and physical activity occurs. As such, space becomes a backdrop to human and non-human life. Crucially this conception treats space mathematically. Studies that understand space in this way seek to map phenomena *onto* space and scientifically hypothesise relations between people, locations, events, and so on, to reach concrete, objective knowledge through patterns, models and formulas.

However, notions of Euclidian space have been challenged by the popularity of the term **place** (brought into focus by humanistic geographers). Where space has been taken as abstract and geometric, place defies such definition. Place is space that has been made meaningful (Cresswell, 2004). No longer abstract, the container that is space, becomes 'filled' with human significance, care and attachment (Tuan, 1977). Ideas of 'place' move geographers beyond thinking of space purely in a locational sense. Studies that focus on place are interested in our metaphorical as well as physical relations with the world. For example, we can be 'in' a place physically, but we can also feel 'in' a place emotionally. This unlocks geography from Euclidean dynamics. Accordingly, geographers who are interested in unpacking human experience of particular locales and also the socio-spatial significance of those relations, often draw on the concept of place (see, for example, Cresswell, 1996).

However, popular today are ways of thinking about 'space' that distance it from Euclidean roots, without opposing it to 'place'. Indeed, another way of thinking spatially is to consider what is called 'social space' (see Massey, 2005). This develops from a **relative** understanding of space. Following the writings of French theorist Henri Lefebvre (1991), in his seminal work *The Production of Space*, we can think of space as **co-constituted** with human activity and non-human actors (objects, nature and so on; see also Massey 1997). Space is only made (or produced) through the relations that form it.

Crucially, an undergraduate project should engage with spatiality. When designing your dissertation, look back over your research question and ask yourself if your project is *spatial* – does it engage space (in whatever way) – to arrive at a *geographical* contribution.

Time and Temporality

In addition to space, time has become an important focus of human geographical research. Spatial relations (as described above) do not occur outside of a second dimension through which life is lived: time. Time becomes relevant in all kinds of ways to geographical enquiry. For starters, human geography projects often explore a given phenomenon, event, place, person and their spatial relations at a particular point in time. My own work on radio broadcasting has, for example, investigated the spatial politics of pirate transmissions between 1964–1991 (see Peters, 2011a). Thus, time is a way in which we *locate* our studies.

Moreover, time is often part and parcel of the spatial relations we explore in our studies. The work of Hägerstrand exemplifies this point. His important research demonstrates that spatial activity is never outside of the confines of time. For example, when engaged in the practice of walking to work, we aren't just moving through space, but we are also moving through time. Thinking of distance alone (i.e. only space) limits how we grapple with human experience. The movement of people is not merely impacted by the space through which they travel, but by the time constraints imposed in doing so. Hägerstrand, then, challenged the purely spatial analysis of proximity, distance and relationality – highlighting how a focus on space alone tells but half the story.

Whilst Hägerstrand is often recognised as bringing temporal analysis to geography, all sorts of engagements with time now shape the discipline. Like space, time can be thought of in mathematical, scientific ways, or in relational, subjective ways. On the one hand, time can be understood as a container in which social and physical change occurs. Accordingly, time is taken to be **linear**; constantly moving forwards. This has resulted in a process-orientated view of space and place where we can map the changes to space and place over time (see Pred, 1984).

However, geographers have also explored what are called non-linear relations between time and space. Scholars examining the spatialities of memory and heritage, for example, often think about the ways in which the past and present fold into one another (see Hoskins, 2007; Nora, 1989). This **non-linear** way of thinking about time is evident in one of geography's most pivotal innovations: time–space compression (Harvey, 1989). In his book *The Condition of Postmodernity*, Harvey (1989) explores the annihilation of distance (of space) through time. In a globalised world it becomes possible to communicate over greater spaces (compressing traditional notions of time) which provides flexibility to, paradoxically, spread labour, goods and services further afield (annihilating distance). This idea, that time can compress and distance can shrink, upsets geometric, measured ideas of both time and space, and alerts us to an image of time and space that consists of folds – a more 'scrumpled' geography (as Marcus Doel puts it, 1996). From this, geographers have begun to think about the ways in which time and space come together in more novel ways, investigating the temporalities of life – the rhythms of activities and the pacings of life (see Mels, 2004).

Not all undergraduate projects will engage explicitly with time, but it is worth appreciating that it is there – and cannot be separated from space. Again, we have seen that there are different frames to thinking about how time matters. When designing your dissertation it may well be useful to consider how time fits in to your spatial contribution and how it relates to the spatial examinations you are making.

Movement and Mobility

What is evident through the two previous sub-sections is that movement is fundamental to the ways in which we engage with space, use time, make place and produce temporalities. Movement and 'mobility' (a term used to describe politicised and power-filled movements) have started to shape and arguably define the interests of geographers who recognise that our world is a mobile world (see also Adey, 2009: Cresswell, 2006; Urry, 2007). Indeed, in a paper featured in geography's landmark journal *Transactions of the Institute of British Geographers*, Peter Merriman expresses the idea that scholars take another concept as important in geographical endeavour: movement (2012). Movement, he argues, is a key way through which space and time are experienced, progressed through, challenged and maintained.

In the past, our relationships with space and place were thought to rest upon stability and rootedness. To know a place, to belong, we must be fixed in place. This was known as a **sedentary metaphysics** (Cresswell, 2006: 27). However, with the development of a 'new mobilities paradigm' (Sheller and Urry, 2006), scholars now recognise that our relationships with space and place are often connected to the ways we move (or are unable to move) as individuals: the movements made possible (or not) via transportation and infrastructure (such as rail, the motor car, airplane and pipelines and cables); the movement of objects (parcels and goods); and virtual movements (made possible via technologies such as email) (see Cresswell and Merriman, 2012 for a good overview, and Burrell and Hörschelmann, 2014; Birtchnell et al., 2015, and Turner and Peters, 2017 for some more particular examples). In other words, mobility pays attention to the dimensions of power that shape the manifold ways in which people, objects and ideas move on a variety of scales (from the individual body, to global connections).

Returning to Peter Merriman then, he contends that **movement-space** is a new way of thinking, precisely because movement cannot be reduced to a particular time or space. In other words, movement is not something that happens *in* time and space, but is part and parcel of unfolding, relative time-spacings and space-timings (2012). If we follow Merriman, movement and mobility are not topics of investigation *within* geography (a topic we study to investigate its spatial and temporal character, politics or ramifications) they are, along with space, place, scale and time – core organising concepts of our investigations as geographers.

Accordingly, to echo the previous two sections, undergraduate dissertations, whether deliberately or not, often consider a world on the move (in various guises). When designing your dissertation it could be useful to ask how the core concept of movement is part and parcel of the geographies you explore. Indeed, identifying your approach to dissertation work relies on also identifying where the geography resides in your investigations. This means taking space, time and movement seriously. In the Graduate Guidance box below, Robert Sheargold discusses how 'ways of knowing' shaped his dissertation work.

THEORY, PHILOSOPHY AND FANTASY FOOTBALL
ROBERT SHEARGOLD

Inspired by a personal interest (and participation) in fantasy football leagues, and a series of lectures on virtual space, my undergraduate dissertation sought to explore the relationship between physical and virtual spaces through research that would involve me building and following a fantasy football team and speaking to a community who competed in online leagues. I would also conduct a textual analysis of the online webpages that hosted the fantasy football leagues.

There was a strong philosophical, theoretical and spatial underpinning to my research project from the start. Before starting this dissertation I had already situated myself within anti-realist approaches to research. I was interested in the individual subjectivities of each fantasy manager and their experiences of negotiating virtual and physical spaces. From this foundation I was able to pick qualitative research methods that were best suited to the collection of data that helped explore the multiple experiences of engaging with online and offline spaces.

Inspired by different geographic theories of space, I began to explore how spaces of the internet were different to physical spaces. I started to find inspiration in the works of key spatial thinkers like Lefebvre and Foucault. Foucault's notion of heterotopia became a way of thinking about how the virtual domain of the fantasy football website mirrored and subverted physical space (see Foucault and Miskowiec, 1986). For Foucault, all spaces are related, and it is these relations that define spaces. However, there are instances where relations both mirror surrounding spaces, and also invert and contradict surrounding spaces. The latter is a heterotopia.

(Continued)

(Continued)

Heterotopia consequentially became the crux in my theoretical framework for thinking about how users' bodies inhabited multiple spaces simultaneously through mediation. Through this theoretical lens I was unknowingly adopting a post-modern way of thinking about the world. While these spatial theories initially seemed a world apart from the study of fantasy football, such ideas helped me map the geographies of fantasy football.

If I was to offer advice on using theory, I would say it is important to remember you are not expected to know everything. Theory can be daunting, but trying to read complex ideas can help in making sense of your own research. I remember for the first month of my dissertation, I felt like I had to consult geography handbooks, dictionaries and my supervisor almost constantly in a bid to understand what I was reading. But these ideas, though challenging, made my project better. I didn't force theory on to my project, rather it emerged and guided my ideas as they arose during the research process. Without theory then, I would have had great difficulty in understanding and mapping the geographies of fantasy football.

Philosophical ruminations: Moving forwards

This has been something of a heavy chapter, but it is a necessary one. When writing this book I grappled with where exactly the chapter should be placed. This is because our approach to geography and our use of theory matter in all stages of research – designing it, doing it and delivering it. Eventually it has ended up here, near the start, in recognition that we begin every research project by thinking about what interests us (and what doesn't). The decision about what we want to study, and how we want to study it, is driven (whether we are aware of it or not) by our 'way of knowing' the world.

Thus, it is useful to identify our approach from the very start – our approach to geography in a broad sense (interrogating what kinds of knowledge we subscribe to, objective or subjective or both); and to more specific theoretical framings (those 'isms' of social theory that are applied geographically). It is also essential to identify how our approach connects to the fundamental tenets of geographical endeavour: space, time and movement. In short, this chapter has been a whistle-stop tour of how we approach our research practice, through 'ways of knowing' and understanding (Aitken and Valentine, 2006: 1). However, the dip-in, dip-out structure of this book means that you can refer back to this chapter when necessary, as your research progresses.

The relevance of the chapter will become more obvious (and applied) as we move into the more practical stage of designing the dissertation in Chapter 3.

It is vital to design research, and plan to use methods appropriate to what you want to find out. If you want to find out about thoughts, feelings, emotions, opinions, values – how people 'make' their world – you must use methods which allow you to uncover subjective meanings. If you want to find figures and numbers and make general claims – believing there is a world out there ready to study – you must use quantitative methods which allow you to show statistical significance or mapped relations. Being a geographer is not simply about a direct, hands-on engagement with the world – it is appreciating that such engagement is shaped by, and in turn shapes, ways of thinking and knowing. As noted in Chapter 1, a dissertation is about the production of knowledge – of ideas. A dissertation makes the reader think. And thinking can be very powerful indeed.

Chapter Summary

- Our engagement with the world is shaped by particular 'ways of knowing'. Thought and practice are always connected because ways of 'doing' research are embedded in ways of 'thinking' about research.

- It is useful to identify our approach to geography, and the type of knowledge we want to arrive at (objective or subjective, or both), as this will shape the focus of our dissertation project. Do we see a world that can only be known through the laws of science; or one that cannot be measured and modelled? Do we think the world is a geometric and pre-existing realm for us to extract data from, or something that is made by us through the attribution of meaning?

- We can often locate our approach by identifying how it sits within and speaks to a particular body of theory (such as humanism, Marxism, feminism, post-modernism and so on). This can help inform the sorts of methodological approaches we take, and frame our interpretations of data, to make sense of it in relation to our research question.

- Our approach to geography should identify with the fundamental, central concerns (or 'basic organising concepts') of geographers – space and place, time and temporality, movement and mobility. The approach taken in an undergraduate dissertation should identify with how the subject under consideration is *spatial* and *spatialised*.

Key readings

Aitken, S. and Valentine, G. (eds) (2006) *Approaches to Human Geography*. London: Sage.

Cresswell, T. (2013) *Geographical Thought: A Critical Introduction. London*: Wiley Blackwell.

Graham, E. (2005) 'Philosophies underlying human geography research', in R. Flowerdew and D. Martin (eds) *Methods in Human Geography: A Guide for Students Doing a Research Project* (second edition). London and New York: Routledge. pp. 8–33.

Shurmer-Smith, P. (ed.) (2002) *Doing Cultural Geography*. London: Sage (particularly Chapter 2).

3

GETTING GOING: FINDING A TOPIC

CHAPTER MAP

- Introducing topic selection
- Topical challenges: A personal story
- Beginning: The thoughtful and creative stage
- Moving on: Finding a research problem

Introducing topic selection

There is something that binds academic staff and students together. All academic staff (your lecturers, personal tutor, dissertation supervisor and so on) will have once been in your shoes. We have all written our own undergraduate dissertations. We have all had to start with a blank sheet of paper when beginning the process of research. Johnston and Sidaway (2004: 2) note that university departments and schools are made up of 'experts' in a given area. Such experts don't simply know a discipline, or an area of it, inside-out. They conduct research that contributes knowledge to that discipline (see also Chapter 1). Yet all experts start somewhere. We have all had to conduct research for the first time. It might be difficult to envisage some of your lecturers and professors as students themselves, but they (we) have all experienced the trials of grappling with research design, application and delivery, having had little to no experience of it.

Finding a topic is the first phase of research design, and it is possibly the trickiest period of the dissertation process. By 'finding a topic' I am referring to the task of identifying a research area and then, more specifically, a detailed focus for study that addresses a research problem. For example, we might enjoy lectures that are broadly in the area of 'sustainability' and decide we want to examine the role of the new carrier bag policies on consumer behaviour as a more specific, targeted (and feasible) question related to this theme. But how do we decide upon the area we want to research and find that more specific focus? Importantly, how do we find a topic that contributes something to an existing body of knowledge, without, to follow Hoggart et al. (2002: 38) repeating old ground that has already been covered? As noted in Chapter 1, a dissertation should offer something towards an existing body of work, contributing to our stock of geographical knowledge. It should address a research problem, offering a means of responding to that particular enquiry.

Of course some students will have an idea for their dissertation quite early on. You may well have an urgent question relating to flood defence management (see Werritty, 2006), studentification (see Smith, 2008) or another area of human geography altogether. But this is rare. And it still requires us to conduct a search of, and review of, the literature to see if the problem has been tackled before. Thus, finding a topic is not a straightforward task. It requires us to move through a process that is **thoughtful and creative**, **academic and informed**, and **practical and manageable**– sometimes several times over.

Let's think about these stages in more detail. The first stage is the idea-generating stage. Here we have to think about the sorts of things that interest us and that could be potential projects. This usually requires us to reflect upon what elements of geography inspire us and the parts of the world that intrigue us. We cover this stage throughout the course of this chapter. The academic and informed stage follows and this is where we investigate how our idea links to the geographical literature and can contribute to it. This stage is intimately connected to the previous one as our ideas are typically informed by the geographical work that we have read already. We cover this stage in Chapter 4, 'The next step: developing your research question'. Finally, there is the practicality and manageability stage. This is where we need to assess if our project is workable. We cover how to ensure your project is 'doable' in Chapter 5.

Accordingly, in this chapter we deal with the broadest part of topic selection – settling on an area of interest. We begin this chapter with an autobiographical account. I recall my own experience of finding a topic, and consider what we might learn from this experience, highlighting some practical (and hopefully reassuring) information for getting started. We then move through the beginning stage of topic selection with a guide for selecting your own dissertation topic. Here we focus on generating broad ideas that can in turn lead to the identification of a distinct research problem (see Chapters 4 and 5).

Topical challenges: A personal story

So we begin with a story. Back in 2005 I was an undergraduate student studying for a degree in Town Planning with Human Geography. I had chosen this degree, following school, after a year working as a civil servant. I wanted to study a subject that led to a profession at the end of the three-year programme. I had planned to be a planner. I opted to take human geography with planning because I'd always liked the subject at school and I thought it would blend well with the study of spatial planning and policy. Unfortunately (or fortunately) for me, I was more fascinated by cultural geography than processes of city and regional planning. For me, this type of geography just 'clicked'. When I came to plan my undergraduate dissertation, I was in a quandary. I still intended to pursue a career in planning. However, I was not sure a planning-related dissertation would really enthuse me.

I started the process of topic selection by trying to identify a broad area to work within. I decided my dissertation would be within the field of 'urban planning' (broadly defined). Starting here, I began to home in on a more specific idea. I had taken several lectures on planning policy interventions to encourage economic growth. I'd found this interesting because I came from Slough and the town had a large industrial estate. I'd spent the early years of my life fascinated by this area, cycling around the huge office blocks, factories and cooling towers. I began to wonder how government and European planning policies worked to encourage business and how that had shaped the growth of Slough's industrial estate. I read around the topic and I found similar studies that used policy analysis as a way of identifying how areas could grow (and how that growth could be limited). I couldn't find anything on Slough specifically. But Slough seemed interesting because of its geographical significance (M4 corridor, half-an-hour from London, a stone's throw from Heathrow airport). Alongside examining policy, I had a contact who worked for the private company that owned the estate. I thought I could speak to them and other relevant persons about the role of policies on business growth. Together with reading, discussions with my personal tutor, more reading, and jotting down ideas, I came up with a main research question and some aims and objectives that spoke to an existing body of knowledge, offering a small contribution to that field (more information on this in Chapter 4).

However, my heart wasn't in it. Tugging away at the back of my mind was a set of lectures on the concept of 'place'. I'd listened, captivated, as the lecturer had charted various ways of engaging with and thinking about the term. How could it be that place was such a ubiquitous but geographically loaded word? How was it that place could be so wholly significant to our experience? How could place be represented, but also performed; produced, but also reproduced through discourse and meaning, practice and action? How was it that place was struggled over and fought for? And how could people feel a sense of pride or belonging, hatred or loathing based on their relationships with place? (see Anderson's 2015 book *Understanding Cultural Geography: Places and Traces* to find out more). I read Tim Cresswell's seminal text *In Place/Out of Place* (1996) (my first foray into reading a book cover to cover) and it was settled. It wasn't in my plan (I was already emerging as something of a poor planner). I backtracked. I changed my dissertation idea. I was going to study something to do with place.

But going back to the drawing board was no easy task. I had to repeat the process of homing-in on a specific area. Wanting to explore 'something to do with place' was not going to constitute a valid research topic. I had to read and read some more.

I had to see if there were gaps in the literature that I could attend to. I had to identify a more specific area around which to explore this multifaceted concept of 'place'. I had to jot down ideas, chat to relevant staff members. I had to find a research problem or issue to address related to 'place' and refine a topic area for examination. In the end, the inspiration came from David Crouch (2002) and his work on the relationship between place and tourism. For Crouch (2002), much of the geographical work on tourism had considered the politics of cultural representation. How were places produced and packaged by the tourism industry in brochures and other texts? Crouch (2002: 207) argued that there was a need to take seriously embodied encounters with places, to fully understand how tourist places are engaged with and also re-shaped through lived practice. At the same time, I had been watching a red open-top tourist bus circling the city, relaying a plotted narrative to those on board. I went to see my tutor.

My dissertation emerged both from the literature and from direct contact with the world. I formed a research question which sought to consider how the tourist industry produced and packaged places through performative techniques on tours. Crucially, I found a topic that was of real interest to me. If I had to justify it (and justify leaving behind my policy-relevant planning topic) – I felt this project mattered too. Often we think that the tourist industry presents us with places *as they are.* In this project I was able to see the power at work in performing a specific version of place on the tours. Places were produced and packaged in a certain way to include some narratives and exclude others through the active performances of tour guides. The histories presented were partial in order to cater for the expectations of tourists (as one company told me). It was an absorbing insight into how something like tourism is far from innocent but has a role in shaping our understandings of spaces and places around us, in an image the industry itself promotes.

Learning from experience

There are a few of things to note from my own dissertation story; things we might learn from my experience. I'll list them for ease.

Select a topic that interests you

Firstly, and perhaps most significantly, it is important when you search for a topic, to select something of real interest to you. If we select a topic that is somehow lacking (in our interest of it, in rationale, in potential contribution to the discipline), it is difficult to sustain the study and produce a relevant and interesting piece of work. As noted in Chapter 1, the dissertation process occurs over quite a long period of time. It is vital that you select a topic that can hold your attention for almost a year. You have to enjoy the reading that contextualises it, and feel excited about the research when you conduct it. Certainly there are parts of all research that are tedious (transcribing audio files of interviews into written form might be one example), but on the whole, the topic should appeal to your concerns and curiosities (see Phillips and Johns, 2012). Reflecting on your interests is the first step to finding a topic. Considering your **identity as a geographer** is a good place to start when thinking about designing, doing and delivering a dissertation. Task 3.1 helps quiz you on the sort of geographer you want to become.

TASK 3.1

Get yourself a pen and piece of paper. Spend 10–15 minutes reflecting on the following questions (it is useful to jot down what immediately comes to mind).

- Why did you decide to study geography?
- What grabbed you about what the subject could offer?
- Which geographical issues interest you the most?
- Why?

Look over your answers. You might find that your *identity* as a geographer is emerging. Geographers often refer to themselves as 'political geographers', 'environmental geographers', 'development geographers', 'historical geographers'. What kind of geographer are you? In turn, what sort of research might you therefore want to conduct?

Ensure your project has a rationale

In spite of our interests some students (like me, initially) select projects that we are convinced are *best* for us to study. We may select a topic area because we feel it is the most relevant or serious – it is the sort of important research that we imagine human geographers conduct. Yet, as this book has emphasised several times already, human geography is an incredibly diverse discipline. All areas we study can be important, even if they don't appear so at first. Studies of the banal and everyday are just as relevant as 'big' questions of global importance (see Holloway and Hubbard, 2001). Look at your responses to Task 3.1 and use this for initial direction in finding a topic. If you know you preferred the lectures and readings related to a course in development geography above all else, then it is no good planning a project that is far outside of this remit.

Make sure you read

My dissertation story reflects the importance of reading. When trying to find a topic it is essential to read – both to find inspiration and to develop your idea from the literature. Reading is an important part of the puzzle for finding and settling on an idea. As Aitken and Valentine note (2006), all research emerges *from* existing literature. From the literature we gain an idea of what research has already been conducted. This allows us to identify what is left to study or what contributions can be made to existing bodies of work. It is important to read contemporary literature to get a sense of what research is being conducted in the discipline currently.

This helps build a rationale for your project. Although today my tourism dissertation might seem quite dated, at the time it tapped into contemporary debates shaping human geography. We will cover reading in greater detail in Chapter 4, but reading was vital to sparking my interest and developing my idea.

Draw on staff expertise

My experience also demonstrated the importance of drawing on the knowledge and experience of staff in the department in which I was studying. On the one hand, staff provided a soundboard. They were able to suggest further readings that would be helpful in honing my ideas. They advised on angles I might take with my project that I hadn't thought of. They offered critique when I was following a dead-end path. Most of all, they encouraged me. When beginning a dissertation it is useful to talk to your personal tutor or to a staff member who is relevant to your potential idea. Ensure you look at staff contact hours and use the appropriate channel to get in touch with them (your university will likely have its own policies on this). Remember, though, that staff cannot find a topic for you. It is *your* human geography dissertation. However, they can help you in making decisions by offering guidance. They can be especially helpful if you change your mind.

Know what to do if you change your mind

The good news is that changing your mind isn't unusual at the start of a project. Settling on an idea or even a rough area of interest can be difficult, especially if there are several areas of human geography that interest you. You might also feel pressure, as some of my students do, to simply pick something, *anything*, just so you have a topic to study. It is best if you are going to change your mind, to do so at the start of the process when generating ideas. This might seem like a strange thing to say. After all, how do you know at the start that you might want to change your mind? What I mean is that it is useful at the start of the project to play with different ideas. Use your notebook (as introduced in Task 1.2) to jot down any ideas you have that might provide inspiration for a project. Play with different ideas at the beginning and test them for originality and feasibility (see the chapters to follow) before settling on the project you will pursue. It is much more difficult to change your mind later on.

Some institutions will have rules on changing your project once you have completed any health and safety and ethical approval forms (see Chapter 5) or once you are a substantial way into your project. Indeed, it is rarely recommended that you significantly change your project once you have a firmly designed project and the research is underway. This is because you will have lost the time you would need to design, do and deliver a new project. If you are in a dilemma it is best to speak to your dissertation supervisor. They will be best placed to offer advice if you are changing your topic of study.

Beginning: The thoughtful and creative stage

In the previous section I considered my own experience as a window through which to consider some of the trials and tribulations of topic selection. But how do we form ideas in the first place? Idea generation is dependent on being thoughtful and creative. However difficult this task is in practice, there are strategies that can come in handy. The best way to begin the creative process of topic selection is to complete Task 3.1 and ask yourself what areas of geography most appeal to you. From here we might consider some of the following to help us find a more particular area of interest.

Engage with academic journals

Journals are the most up-to-date publications relating to a given discipline. Unlike books, journals are published more regularly, so they feature cutting-edge research on the most pressing questions shaping debates in human geography. If we are seeking to find out what the 'state-of-the-art' of human geography is, we should turn to journals (see Information Box 3.1, below). These can be a good starting point for inspiration when devising a project. Recently a student of mine was stuck for an idea. He completed a task similar to Task 3.2 below and came across an article about expatriate communities and sense of identities (Bochove and Engbersen, 2015). This appealed to him because he had relatives living as expatriates in Singapore. From this paper he found more readings via the reference list at the end of the paper. This literature inspired him and would later become the foundation of his dissertation work.

However, whilst Task 3.2 is helpful in finding an area of interest, it doesn't entirely help us in finding a research problem – a specific question we want to ask. Thus, when perusing abstracts, and then reading articles, you should bear in mind whether the topic requires any further investigation. Are there any fresh lines of enquiry to be followed or new angles to consider? As Gatrell and Flowerdew note (2005: 40), journals can be especially helpful in this respect as they often outline – generally at the end of the article – where work in a given area should progress in the future. When making notes on reading, then, bear in mind if there is a contribution you could make through your own research, to the debates you are reading about.

TASK 3.2

Allocate yourself a couple of hours in your schedule and select one geography journal. It is a good idea, especially if you are stuck in finding a particular area of human geography of interest to you, to select a journal that publishes in all areas of human geography. Landmark journals such as *Progress in Human Geography* (see Go online box below) or *Transactions*

(Continued)

(Continued)

of the Institute of British Geographers are particularly useful publications to consider. *Area* offers short, incisive articles on new research topics. All institutions subscribe to different journals, however, so find one that your university holds (either in paper copies or online). If you know a particular area of geography you are hoping to focus on already, you might look at more specific journals (for example, if you are interested in urban and regional geographies you might look at *Environment and Planning A*, or if you have been inspired by cultural studies, the *Cultural Geographies* journal. See Information Box 3.1 below for a brief list of major journals and their areas of concern).

Get a notepad and pen or a new computer file and take a look at issues from the journal over the past two years. This will give you a sense of recent work being conducted in the discipline. Find three articles that appeal to you. To do this, browse article titles in the first instance. Is there anything that grabs your interest? Next, read the abstracts of interesting articles to gain a sense of the arguments and the debates attended to. What is it about specific articles that is of interest to you? Make notes on these questions to try and identify what topics, debates and subject matters are appealing to your curiosities.

Of the three articles, next consider which abstract you enjoyed most. Read the article. Follow this up by looking at further readings indicated in the reference list. Here you might find related articles that are of interest.

Go online! Visit **https://study.sagepub.com/yourhumangeography** to access Sage articles from the journal *Progress in Human Geography*. This collection of articles showcases some of the wide-ranging, cutting-edge work in the discipline. Have a browse. Do these articles provide any inspiration for your own human geography dissertation?

INFORMATION BOX 3.1
KEY JOURNALS

Listed below are some key human geography journals you may wish to consult in generating ideas. You can also Go online! to access a list of web links that will take you to the home page of each of the journals noted below.

Annals of the Association of American Geographers – publishes empirically informed articles in all areas of geography

Applied Geography – publishes work that applies theory and methodology explicitly (i.e. GIS, statistics and so on) to understanding socio-spatial relations

Area – publishes short articles on new and cutting-edge research in all areas of geography

Cultural Geographies – articles explore the cultural dimensions of environment, landscape, space and place

Economic Geography – features research that attends to significant economic geography issues around the world

Environment and Planning A – interdisciplinary journal that focuses mainly on urban and regional research

Environment and Planning D: Society and Space – publishes philosophically and theoretically driven work that explores relations between society and space

Gender, Place and Culture – focuses on research relating to gender and feminist issues along with work that explores the intersections between gender and other facets of identity

Geoforum – publishes wide-ranging research across environmental, political and social geography that explores relations with space and time

Geography Compass – publishes high-quality review articles on key areas of investigation in geography. There is a 'political' compass, 'cultural' compass, and so on for each area of interest

The Geographical Journal – publishes policy-relevant and informed articles that have distinct 'relevance' to the wider world

Global Environmental Change – articles consider the human and policy implications of environmental change across multiple scales and times

The Journal of Historical Geography – publishes on all areas of historical geography: the histories of geography, and historical geographies of socio-spatial relations

Political Geography – articles focus on the spatial dimensions of politics, and geopolitical relations

Population, Space and Place – publishes empirically led and theoretically informed research on understanding population

The Professional Geographer – publishes short empirically informed and incisive articles in all areas of geography

Progress in Human Geography – publishes the most cutting-edge work in human geography; papers indicate where and how geographical study can progress in the future

Social and Cultural Geography – articles focus on the spatial dimensions informing both social and cultural life

Tourism Geographies – papers consider tourism and leisure from a geographical perspective

Transactions of the Institute of British Geographers – a landmark geography journal publishing high-quality papers across all areas of the discipline

Urban Studies – the leading journal for debates concerning urban spaces and places; the urban condition, urban policies and planning; and regional and global connections

Visit the library

With many journals now available online through your university account – and for that matter e-books too – we can often forget how useful a library can be. For a start, immersing yourself in the library can be beneficial for putting you in a good state of mind for the dissertation and the academic task you are undertaking. Moreover, simply browsing interesting titles can be a good starting point for finding a topic idea. As Gatrell and Flowerdew note (2005: 40), 'there is much to be gained from browsing the library shelves'. When we 'look' for topics online, we are reliant on having a starting point – a term, a topic, an event – which we can type into a search engine. When we browse for books online, again we must have a sense of the kind of book we are looking for so we can indicate this in the search box. Unlike web-based search engines, searching the shelves of a library permits us a greater freedom to stumble across titles we might not have otherwise found. You will notice much more by browsing shelves than browsing an online library catalogue.

Start in the human geography section. Scan the bookshelves for titles that stand out, and make a note in your diary (see Task 1.2) of any potential areas that spark an interest, which you could later follow up on. Take note of particular authors who seem to be writing work which you enjoy; you can also follow up these names later too. If you have already read something that is of interest to you (geographies of retail for example, or global–local relations), go to the shelves that surround the book you've engaged with and see what else is shelved in that location. You may find other work that is in a similar area of interest. You might also want to cast a wider net. Certainly, it is important that your dissertation topic is geographical (see Chapter 2). But you may also gain inspiration from browsing the stacks of books related to cognate disciplines too. For example, if you know you are interested in gender relations, why not spend some time looking at titles in the sociology or women's studies sections, for inspiration?

Reflect upon the lectures you have attended

I have suggested browsing journals and books as a first approach for finding a topic, because, as we know, our topic has to contribute to an existing field of knowledge in human geography. It is easier to find not only an area of interest, but a research problem related to it, if we have a sense of the academic literature that already exists. However, we can also refer to lectures we have attended for inspiration too. On a very broad level, lectures we have enjoyed signal an area of interest we might pursue for a project. A student I was recently supervising was inspired by a set of lectures on geopolitics and the media (and by an engagement with Klaus Dodds' excellent work on James Bond, 2006). Motivated also by an interest in oil and resource crises from another lecture series, they wondered at the potential of a project considering how the media used popular films to depict resource wars. Since much work had already been conducted on oil (see Bridge and Le Billon, 2013), they found a gap related to diamond mining and worker and resource exploitation (also following Le Billon's work, 2008). They sought to explore the politics of representation of this crisis in the media. This, then, became their project (more honed of course, with a distinct question, aims and objectives, and underpinned by relevant reading). The inspiration, though, came from a collection of lectures.

If you have been inspired by a lecture yourself, look at the reading list provided and follow it up with reading. Are there any gaps for further work in this area? In addition, as Gatrell and Flowerdew suggest (2005: 42), you could speak to the lecturer who has delivered the course (as noted earlier). If you have a rough idea of your interests, discussion with them may help you home in on a more specific focus and a distinct research problem to address.

Consider contemporary issues and current affairs

Geography, as we know, is concerned with the world around us and how space, place, time and motion relate to our experiences in the world. You would be hard pressed to find news stories of current concerns and affairs that were *not* geographical. Geography matters (or can matter) to a host of topical issues. Indeed, even subjects that don't appear geographical can probably be thought of anew with the application of a *spatial* lens. For example, repeated news stories on binge drinking and alcohol-related crime might not seem like the basis of a geography project. However, one study has questioned how street pastors and charities work in the night-time economy to mitigate the issue of drink-related accidents and crime. In particular they have sought to understand how religious organisations work in what they call 'post-secular' – or non-religious *spaces* (see Middleton and Yarwood, 2015). In addition, Gatrell and Flowerdew (2005: 41) argue that local news stories can be particularly inspiring because they are close to us and of direct interest. Local news stories can also be helpful in formulating a feasible project as we can often easily contact the relevant people, and access the appropriate documents, local to the area where we are based.

That said, we should exercise some caution against drawing on very recent events as inspiration for a human geography dissertation. A recent news story may be hard to examine for a dissertation because the 'issue has yet to run its course' (Hoggart et al., 2002: 46). Indeed, we could formulate a research problem to respond to in relation to a new policy or a new building development, only to find that one (or the other) falls through. Then, as Hoggart and colleagues note (2002: 46), 'disaster' ensues. You will be left with nothing to research. Often with emerging issues there can also be a lack of data to collect. Thus, we can use the news broadly for inspiration, but avoid very recent issues.

Consider your own personal interests

Many students, when trying to settle on a dissertation topic, turn to their own interests for inspiration. This might be because we are aware of a pressing matter, and it matters to us. For example, recently in Aberystwyth, students have protested about maintaining exclusive Welsh-language halls of residence. This has inspired a glut of projects exploring the geographies of banal nationalism (Bilig, 1995), language-use and protest. In addition, many students are driven to explore geographical concerns based on their own beliefs and values. For example, feeling particularly strongly about food miles and carbon footprints might lead to a project that is sensitive to these concerns. I recently read a very good dissertation that focused on harmful emissions caused by refrigeration, where the student noted their own ethics as a rationale driving the study.

But we do need to be careful when focusing on issues of personal importance. For a start, we can be blinded by the relevance and rationale of our project, simply because it is of interest to us. Just because we are engaged with a subject doesn't mean it is worthy of study. We should always combine the development of our topic ideas with academic reading. Does your idea speak to/with broader disciplinary debates? Can it help us understand those debates through application to a new empirical case study? We also need to be wary of our own personal involvement in the project. Studying something we are close to can bring benefits in accessing the people, documents and places that we might require to answer our research question. But it can also mean our perspective is clouded. There is a vital need to be reflexive when conducting research in areas we are personally involved in (see Heley, 2011; Rose, 1997; and also Chapter 6). We might even ask, as Hoggart et al. do (2002: 47), if we are the appropriate and best person to even investigate the topic.

Finally – it can be tempting to focus on an area of interest to us because it seems somehow easier. Students sometimes pick topics related to sports societies or social clubs they are involved in as it seems an obvious choice. There are two cautionary notes to make here. Firstly, such topics, as noted above, might lack a clear contribution to geographical debates. Secondly, a project driven by personal interest could lead you to despise the very activity you enjoy during time-out from your academic studies. A friend of mine from university once told me she bitterly regretted her degree in English Literature as it 'ruined' reading for her. No longer could she sit down with a novel and simply enjoy it. Her brain had been taught to interrogate everything she read. Likewise, using one of our passions as a source of research inspiration can change how we feel about it – so be aware before you commit to a project based on your personal hobbies and pastimes.

As this section comes to a close, Rachael Squire provides some insights into the process of generating an idea in relation to her recent research project. See what inspired her in the Graduate Guidance box below.

GENERATING IDEAS AND FINDING SUB-SURFACE GEOGRAPHIES
RACHAEL SQUIRE

Before embarking on my dissertation I found myself questioning how I would be able to make any sort of original contribution to geographical knowledge. To me, geography felt overcrowded and I felt overwhelmed by the task of intervening in such a busy and thriving discipline. When I began reading about the sea, however, this began to change. I found the sea fascinating and it presented me with an opportunity to explore and immerse myself in an emerging field of human geography. There was and is so much to be discovered about the earth's watery spaces and I found these unknowns to be quite liberating when finding a dissertation topic.

During this time, I stumbled across a great case study on the BBC News website that combined my interests in the sea with those of political geography. Spanish Police divers had been photographed unveiling a Spanish flag over an artificial reef in British Gibraltarian waters causing outcry in both Gibraltar and Britain. I was intrigued by this use of depth for political gain and this led me to Stuart Elden's paper on 'Securing the Volume' (2013) in which Elden makes the case for a three-dimensional understanding of territory – in other words he argues that we need to stop looking across landscapes and instead look beneath, up, down, and around them. I plugged into the debates taking place around Elden's paper and read related work on the topic. From this I was able to identify a small hole that I could fill through my dissertation research.

Discussing my ideas with my tutor and with other students was also crucial to pinning down a topic, as was finding a big piece of cardboard and creating a mind map of my ideas. I had to learn to be comfortable with an inherently uncomfortable process whereby my topic was changing, expanding, shrinking and escaping the box I had put it in at the beginning. Indeed, whilst the sea proved to be the initial hook I ended up engaging with Gibraltar's seas, skies, land, and economy. It was the result of a long process of learning what I was interested in, reading and re-reading literature, discussions, brainstorming and putting pen to paper. It was not straightforward and there was many a time where I felt a bit lost but this was all part of the process of turning my ill-defined interest in a particular space into a well-defined research topic.

Moving on: Finding a research problem

In this chapter we have begun the process of identifying a research problem by considering how you identify a topic in the first instance. As previously noted, it could be that you have a very specific research problem in mind that you hope to explore. However, it is more likely that you will want to contemplate a few areas of interest before settling on a topic to pursue. It might also be the case you are simply lost for where to begin. The first stage of research is to identify a broad area in which we hope to work. Once we have established a broad area to focus on, we must narrow this down to a distinct topic, identifying a specific research problem to attend to in our research. As Gatrell and Flowerdew tell us (2005: 43; emphasis in the original) 'you should always be looking for a research *problem*: a question or set of questions that are worth asking, an issue that merits attention or requires solving'. We attend to how we go about reviewing literature and identifying research 'problems' in the next chapter. This is alongside a consideration of how to form questions, aims and objectives. But for now, start generating ideas. Don't be afraid to follow your thoughts and to be creative. If geography is everywhere, inspiration for your topic is, too.

Chapter Summary

- Finding a research topic relies on the overlapping processes of being creative with your ideas, reading around the subject for academic relevance, and testing for feasibility. This chapter has covered how you formulate ideas from a blank sheet of paper.
- Staff and students all face the dilemma of starting research projects for the first time. For students it is crucial to select an area of study that can hold your attention and interest for the duration of the dissertation project. Whilst some students know immediately what they want to explore, and have a distinct question in mind, most do not. Start broadly and home in on your interests gradually.
- We can generate ideas for a project in several ways. Academic reading is the best strategy as it can help us identify gaps in research, the need for further research, or new avenues of investigation. Lecture series, current affairs, and personal interests can also be source of inspiration, but ensure these tap into geographical debates and aren't selected on a whim.
- Topics should have a rationale – a reason they matter. Ask yourself if your topic area has an academic and/or empirical motive that makes it necessary to examine. Research should address a problem or issue, offering a contribution that adds to debates relating to that concern.

Key readings

Clifford, N., French, S. and Valentine, G. (2010) 'Getting started in geographical research', in N. Clifford, S. French and G. Valentine (eds) *Key Methods in Geography* (second edition). London: Sage. pp. 3–15.

Gatrell, A. and Flowerdew, R. (2005) 'Choosing a topic', in R. Flowerdew and D. Martin (eds) *Methods in Human Geography: A Guide for Students Doing a Research Project* (second edition). London and New York: Routledge. pp. 38–47.

Hoggart, K., Lees, L. and Davies, A. (2002) *Researching Human Geography*. London: Arnold. (particularly Chapter 2).

Kitchin, R. and Tate, N. (2001) *Conducting Research into Human Geography*. Harlow: Prentice Hall (particularly Chapter 2).

4

THE NEXT STEP: DEVELOPING YOUR RESEARCH QUESTION

CHAPTER MAP

- It's question time
- Reviewing literature
- Literature reviews
- Formulating your research question

It's question time

If you were to ask a novice, the subject of Geography seems to be littered with questions. Popular quiz shows and board games present our discipline as one which is concerned with questions of metrics: how large is the population of Aberdeen? How long is the river Rhine? How tall is the highest building in London? What is the GDP of Belfast? As a student of human geography, however, you will be well aware that geographers ask an array of questions that aren't only factual, or concerned with answers that might be simply 'right' or 'wrong'. With an interest in connections between people, space and place, human geographers can ask a host of questions (both statistical and discursive) which seek to 'get at' the shape of our social, cultural, political and economic worlds.

Research of any kind requires questions. Indeed, if the act of research is investigation of a topic to discover something as yet unknown (see Information Box 1.1, p. 2) then it makes sense that questions precede enquiry. However, once you have found a broad topic area that interests you, how do you find a distinct research problem or question to respond to and answer in that chosen field?

Figure 4.1 The egg timer or wine glass analogy; all dissertations are wide at the start, becoming narrower in the middle, before broadening out again at the end.
(Source: Creative Commons, 2015 http://uploads.wikimedia.org/wikipedia/commons/2/2a/1328101886_HourGlass.png; Jorge Barrios (via Wikimedia Commons) http://upload.wikimedia.org/wikipedia/commons/thumb/7/75/Copa.jpg/90px-Copa.jpg

It is useful to visualise the dissertation process like an egg timer or a wine glass (Figure 4.1). All projects begin broadly – like the top of the egg timer or wine glass (whichever analogy you prefer). Projects then become narrower as you settle on a more specific research topic and then a specific question that is the central focus of your dissertation (illustrated by the mid-point of the timer, or the stem of the glass). Your project then widens again at the end, as your distinct project is used as the basis for suggesting a set of further, additional questions that could follow on from your study.

Defining the 'middle part' is crucial for developing a good dissertation project. It is desirable for your project to become narrower because it makes your research more feasible to conduct. Broad projects that ask 'big' questions are often unworkable because they cannot be completed by just one researcher (you) in the limited time available (usually less than a year of a degree, alongside other modules or courses). Accordingly, developing a specific question goes hand-in-hand with designing a practical and manageable study. We discuss such matters in Chapter 5. But for now, we need to grapple with how to turn a broad area of interest to a discrete project with a clear research question.

In the previous chapter, we discussed how you might begin generating ideas, identifying a topic area for a potential project. This was the creative and thoughtful stage. The next step is to develop your idea into an explicit question. This process – of narrowing down your topic – requires a more academic and informed approach. In what follows we investigate how our ideas are connected to the geographical literature and how, in turn, reading can help in forming a research question. This chapter begins by considering how to find literature to review in a given field of interest. This is followed closely by a consideration of how you engage with literature to develop your question. The chapter then culminates by demonstrating how literature enables us to formulate research questions. Here discussion focuses on the fundamental task of writing research questions, aims and objectives.

Reviewing literature

Reading is crucial to a successful research project. When we read, we tend to start broadly (like the egg timer or wine glass) and then home in on more specific readings, searching out a research problem. It is easier to find a worthwhile research problem, or question, if we understand the field of work in which our project sits. By reading we can gauge the work that has already been conducted, and identify possible gaps to attend to, or fresh angles to explore. As such, reading is time well spent at the beginning of your project.

> Go online! Visit **https://study.sagepub.com/yourhumangeography** to access Sage journal articles that document how the process of reviewing literature is vital to formulating research questions, as well as articles that provide guidance as to how to write a good literature review for your dissertation.

But where do we begin finding academic literature relevant to our study? Take your topic area as identified in Chapter 3 (Task 3.1) and from there pursue the following avenues to find relevant readings:

Ask your supervisor

A great place to begin to find relevant readings in your area of interest – especially those essential, 'classic' readings on the topic – is to ask your dissertation supervisor. Supervisors will usually point you towards readings they believe to be useful and relevant to you. However, it is important not to expect your supervisor to provide a comprehensive list of readings. As Flowerdew also notes, sometimes supervisors will not provide whole references, but simply direct you towards particular scholars, areas of debate, or journals (2005: 50). Thus, 'detective work' may be required so you can find the particular reference that is relevant. Ensure you go to supervisory meetings with your notepad/diary (see Task 1.2). Be ready to write down any key journals, article titles or authors they suggest (and don't be afraid to ask for the correct spelling of names they mention). In addition, you can use any suggestions of key readings provided by your tutor to guide you to further relevant readings (see below).

Use reading lists

Lecturers and supervisors can also be a key source for finding relevant readings from the reading lists they (we) compile for any given lectures. If you have been inspired to study a particular topic as a result of a module you have taken, the accompanying readings can provide a useful starting point for grasping the major discussion points that scholars have addressed in that given area. They might also provide readings related to a specific area that really grabs your attention as a possible area of concern. For example, a former student of mine enjoyed a lecture on embodied engagements with landscape through walking (see Edensor, 2010; Wylie, 2005). The reading list helped the student to locate this work in a broader set of debates concerning geographies of the body (Longhurst, 1997). Through reading this work, they wondered if there was the potential to explore types of embodied experience aside from walking. They were particularly interested in kayaking as an activity that they engaged with in their own time. They were intrigued as to how kayakers experienced the water and how this engagement might be different from other kinds of embodied practice (see Anderson, 2014). They then used additional methods of searching literature to uncover more specific, related work that was connected to this idea (see below).

Consult dictionaries and textbooks

The *Dictionary of Human Geography* (Blackwell-Wiley) and other reference works such as the *Encyclopedia of Human Geography* (Sage) can act as useful starting points in identifying key works that have shaped debates. After explaining – in brief – a particular term (for example, 'territory'), they then provide a list of references that can direct you to the crucial literature which foregrounds your study area. It is important, however, not to rely on dictionaries or textbook entries as the only source of literature you use. Whilst they explain things clearly and often in language that is more straightforward than the sources they refer to, it is beneficial to engage with the work of the geographers mentioned first hand. If the purpose of a literature review is to demonstrate how your research connects to existing literature, and then contributes to that literature, it is vital you have a clear handle on that literature directly.

Read review articles

Whilst journal articles present new geographical findings based on research conducted, review articles (not to be confused with book reviews, which are singular reviews of new publications) explain the state-of-the-art of a particular field. They are helpful as they tend to cover a particular area in great depth, highlighting key readings, but in doing so they also make suggestions for future directions of study in the field. This is where review articles can be really pivotal. You can find review articles in several places. Often they are located alongside standard research articles in a journal, or they are placed in a separate section. Other journals, however, deal purely with review articles. *Geography Compass* articles are becoming increasingly important resources in offering reviews of geographical topics. The journal – which is wholly online – is split into easy subsections (Political Compass, Cultural Compass, Urban Compass, Development Compass and so on). The array of topics covered is incredibly broad, so you are likely to find a paper that relates to your interests from which you can both hone your ideas and gain a sense of the literature crucial to that area of the discipline. Whilst Compass articles have their merits, *Progress in Human Geography* is another key journal to consider. Articles here are not traditional review articles, but rather aim to set future agendas. As such, the papers published here have become important benchmarks in defining debates. Accordingly, older articles are usually important context papers and newer papers can assist in outlining key avenues that we might want to consider now.

Go online! Re-visit the **https://study.sagepub.com/yourhumangeography** pages for Chapter 3. Here you will find links to articles in the journal *Progress in Human Geography*. Click on a paper that interests you from the list and read the article. Ask yourself the following questions: What field does the paper seek to review? What are the key areas that define the topic area considered? What new directions are proposed by the author(s)? Does it raise any particular research problems that require investigation?

Use reference lists and bibliographies

The reference list or bibliography at the end of an article, book, or book chapter we have read is a good way of identifying other literatures in a similar area. This helps us to build an extensive collection of readings relevant in contextualising our study. However, as Flowerdew notes, we should be cautious in relying on reference lists alone – especially as a source of up-to-date references (2005: 52). This is because the reference lists of papers will always refer to papers that have been published earlier than the one you are reading. Therefore, whilst the paper itself might present cutting-edge ideas, it will be in based on, and situated within debates that have opened up the space for that particular contribution (rather than the most recent articles that have emerged since). Thus we can use these reference lists as a guide but should be aware they will not provide us with the most contemporary work being carried out by geographers.

Search a citation index

Citation indexes work by providing a catalogue of journal articles, conference proceedings, workshop chronicles, and so on, organised by citations. In other words, authors working in cognate areas will draw on each other's work and cite each other's papers. The index connects these works through a database that will provide a list of all papers that cite a particular article (see Heley and Heley, 2010). Citation indexes help us to find the latest articles and also get an idea of the most pivotal articles (which tend to be the ones most cited). The main citation index used is the ISI Web of Knowledge (WoK). Within WoK, the Web of Science index often proves most useful for human geographers (especially the Social Science Citation Index). It can often appear easier to use a web-based search engine (see below) as opposed to Web of Knowledge, which is more complex to use. However, WoK is a fundamental database for finding literature for your human geography dissertation. Access to WoK is usually via your institution, using an ATHENS password. Once you are logged in, it is worth browsing the database – using a variety of key words (from topic areas to author surnames) – to gain a handle on how it works.

Use a web-based search engine

It is now common to search for literature using readily accessible search engines, such as Google Scholar. It is worth mentioning, however, that search engines function somewhat differently to citation indexes, such as Web of Knowledge. Most internet searches will filter results based on 'keywords' and will match the most relevant articles to search terms. For example, if you typed in 'geography', 'resilience', 'disaster' and 'volcanoes' to a search engine – you would receive a rank order list of articles and resources that have a high number of 'hits' of those terms. In other words, unlike Web of Knowledge, you will not gain a filter of the most cited or important articles, and you may even find papers which are not peer-reviewed, and therefore not of the highest academic standards (see Task 4.1). You can, using Google Scholar in particular, filter your results by year (sifting out older articles) but, even with this capacity, such resources, valuable and easy as they are to use, should be approached with caution. It is useful, when using search engines, to vary your search terms. A search of 'music', 'geography', 'Wales' might bring up totally different results to a similar search of 'music', 'space', 'nationalism'. Do not assume there is no literature on your topic if your searches do not retrieve any relevant results. In short – using a combination of search methods for locating literature is the most thorough approach.

Go online! Visit **https://study.sagepub.com/yourhumangeography** to find web links to sites that help us to search for literature in the ways outlined in this section of the book.

TASK 4.1

When we search for literature that will both inspire our dissertation topic and help us develop a specific research question, we also have to be sure to use sources we can trust. See below the list of sources that we can consult when reading for our dissertation. Jot these down and next to each one consider the benefits of using the sources, and whether there are reasons we should use such material with caution.

Peer-reviewed journal article

Academic book chapter

Wikipedia

Geographic dictionary entry

News article

Monograph

Academic blog

Popular non-fiction book

USEFUL TIP

When compiling reading (using the methods outlined above), save your references in a Word document as you find them. Accordingly, you will begin to develop your own tailored reference list of sources relevant to your project. Moreover, don't just cut and paste these, but format them into a consistent style (consult your student handbook for details.) Eventually, this list can form the basis of your reference list in the final dissertation. However, you need to make sure the final list only includes the readings you've used in the main document (see also Chapter 12).

Literature reviews

So far we have discussed how you find literature relevant to the area you hope to study. But how do we use literature to actually find a research question? This all depends on whether your ideas for a project still require 'pinning down', or whether you have a good idea for a topic. Let's consider these two situations in more detail.

Your current ideas are quite broad still. You have a vague idea of what interests you, but you need to hone these concepts further

When reviewing literature, it is advisable to start by 'casting the net widely' as Flowerdew puts it (2005: 49). This is often an obvious place to start if you only have a rough area of interest and are still trying to pin down your exact study. For example, if you know you are interested in geographies of consumption and material culture you might begin by finding the landmark publications that have discussed this topic in the discipline, using the methods suggested above, before narrowing your search to find more specific journal papers, book chapters and so on, which explore these geographies in light of particular examples and case studies. This might then give you more specific ideas that lead to the identification of a distinct geographical project.

For example, a student at my institution had been inspired by a historical geography module and, in particular, a lecture on the role of museums in selecting and representing history (see, for example, Geoghegan, 2010). However, the student in question couldn't find a question. Surveying the literatures that investigated the geographies of museum space alongside the wider literature in Museum Studies helped greatly. They were particularly intrigued by readings that explored how space is used to route visitors around museums. They were then reminded of a museum close to their home town – a so-called 'Living Museum' – where history was performed in a historic village by actors in costume (as opposed to a museum full of cabinets of curiosities – see M. Crang, 1994). They wondered how spatial practices of museum curatorship and the routing of visitors played out in this type of museum.

Thus, gaining a firm handle on the reading alerts us to ideas which we might then apply to a new case study. It also alerts us to gaps. Another student was inspired by Ben Anderson and Peter Adey's work on civil contingency planning and preparing for emergencies (2011; 2012). They enjoyed this reading but could not see how to establish their own study from it. They began to look for 'gaps'. They soon noticed that much of this work was concerned with managing emergencies in cities (on land) or those that transpire in the air (through volcanic ash clouds and so on). They identified that this work in this area had not explored the sea – or water spaces. Their interest and knowledge of this literature helped them find a niche to explore. Their dissertation eventually went on to focus on the role of life-saving agencies in preparing for emergencies in water.

Your current ideas are very specific. You know exactly what you want to study, even if it isn't yet framed as a question

If you have a quite specific idea to begin with, it is worthwhile broadening your search of the literature. For example, you may have been inspired by Jason Dittmer's work on comic books as windows for exploring geopolitical relations (see Dittmer, 2005; 2012). We might assume, therefore, that this is the *only* relevant literature for a project that likewise considers such comic books. However, reviewing the literature is still necessary. For a start, as noted in Chapter 2, geography dissertations should ideally be situated within the theoretical or conceptual framing that inspires the work. In other words, even if we have a discrete, well-defined idea, we should connect it with broader disciplinary debates. This helps to demonstrate our mastery and expertise of the field we are researching (a key skill of a dissertation study). Secondly, we should always review the field thoroughly (Parsons and Knight, 2005: 26) as we might find that someone has already completed a similar study. Acknowledging this is useful as it shows you are aware of the field, as well as allowing you to develop your idea (perhaps by applying the ideas to a different case study, or speaking with a new group of people) to give your project originality.

Reading is vital in the process of research design. But how we read also matters in developing your question. In the sub-sections to follow we discuss how you review readings in order to extract the key points, which can help you find your own research problem to address or question to pose.

A question of evaluation

It is vital to evaluate what we read. As Kneale tells us (2011), making effective notes is crucial when reading. I have a computer file (titled 'readings') where I make three brief bullet points about each journal article or book, after I have read them. I then have a clear record of the reading that I can return to, which evaluates the text. My three bullet points cover (1) content (the key point(s)/arguments of the piece); (2) context (the set of debates it states its contribution to); (3) critique (any thoughts I have on the piece and if I think the debates could be extended, changed or considered in a new light). **Applying the 3 C's** to any reading you do can enable you to clearly evaluate a piece of literature and actively consider how it might inform your ideas.

A question of problem identification

Secondly, in evaluating readings we can identify potential questions we might ask in our own dissertation. Parsons and Knight, albeit writing from a more scientific perspective, offer a good overview of how we identify topics through reading (2005: 45). Information Box 4.1 below borrows from their overview.

INFORMATION BOX 4.1
LOCATING A PROBLEM IN THE LITERATURE

When reading, consider the following scenarios. They can assist you in carving out a distinct research problem to investigate.

Scenario 1: 'It appears that nobody has investigated this topic – I'll have a go.'

Once you have read around a broad area you might find an empirical area of study, e.g. a particular protest group, a specific urban development, a new environmental policy that is yet unexamined. You might also realise that you can say something new about a theoretical area of concern through your own examination.

Scenario 2: The key authors have 'investigated this topic and raised the question regarding the role of X'; I'll be the one to explore that question.

As noted earlier, many academics will flag areas they believe are worthy of further study – either in review articles or at the end of research papers. We can find questions for our own project by looking to the areas such authors highlight as requiring attention.

Scenario 3: I read an interesting study that explored geographical ideas at location X – I'll see how these ideas compare with ideas elsewhere.

A solid dissertation can offer a contribution by applying existing ideas to a new case study – a different location, an alternative social group, or another political, economic or environmental situation that has as yet to be investigated.

(Continued)

(Continued)

Scenario 4: I came across a reading about a particular phenomenon – 'I wonder if things have changed since then?'

You may be able to establish a feasible study by considering change over time of a particular geographical topic. That said, we should be wary of projects that seek to ask broad questions of how spaces, places, cultures or politics have changed. Rather, these kinds of projects should be well defined and have available source material for study.

Scenario 5: Since the previous study on the topic 'a new dataset has become available' – can this information shed new light on socio-spatial relations?

Studies that statistically correlate, analyse and map socio-spatial relationships are always vulnerable to new data. When new government data (such as a Census or household survey), or new map data (raster or vector) is released, it can provide human geographers with the opportunity to ask established questions which were highlighted in the literature from a fresh perspective.

(Adapted from Parsons and Knight, 2005: 45)

A question of how much is enough

There are many ways we can use readings to find our question. But we might also ask (and I am asked often) how much reading is enough reading? It is difficult to put a number on how many readings you need to complete to have effectively reviewed the literature. The volume of reading you conduct will depend on your project (which will be different from someone else's project) and will also depend on how much you want to stretch yourself academically. In many ways thinking in terms of numbers is unhelpful. Rather we should read until we have adequately – and deeply – understood, contextualised and developed our question. To do this properly will always require more than a mere handful of readings.

Formulating your research question

So far, this chapter has provided ideas for how you use readings to narrow down your ideas and find an academically informed area of enquiry. However, sadly, it can't find that distinct research problem for you. Designing your human geography dissertation is certainly one of the most challenging (and time-consuming) elements of the research process. It can be frustrating (if you can't seem to settle on what to study). It can be overwhelming (if you can't find an avenue to explore). It can also be easy to give in and put off the work required. Yet knowing that this stage is hard should be reassuring. Your anxieties will not be yours alone, but are likely shared with students in a similar situation in your own university, and other institutions. Therefore, stick with it. Keep asking questions – asking yourself what interests you (Chapter 3) and asking what gaps, necessities for further study, or varying enquiries could be examined by reading the literature that engages you (as covered in this chapter).

However, once you have worked through these stages and have a clearer idea of a research problem, it is necessary to turn it from a possible question (in your head, or in jotted note form) to a firm research question that you can include in a research proposal (which your institution may require) and in the final dissertation (which you will write up). In this final section we consider how you develop your question, as well as running through some tips for framing your question to accurately reflect your study area.

Developing the question

Once you have found a potential research problem (using Parsons and Knight's guide, Information Box 4.1, above) it is useful to arrange a meeting with your dissertation supervisor. This will allow you to check that you are on the right track before investing all of your energies in the project at an early stage. Your supervisor will likely raise further questions for you to consider. They may also raise concerns. Task 4.2 aims to help you to define the very purpose and rationale of your study to present to your supervisor. It also helps in developing your *actual* research question.

TASK 4.2

From generating broad ideas (Chapter 3), to searching the literature for a distinct problem to address, you should now have a dissertation topic in mind. The next step is to try and articulate this topic succinctly and clearly, in a way that expresses the purpose and rationale of your project.

Work through the following sentences in respect of your project. Try to explain your research in just one sentence. This might mean writing, editing and re-writing your responses more than once. It is rare to write your research question on the first attempt. Rather, research questions require careful honing to convey exactly what you mean, and what you intend to examine. So be prepared to scribble down your answers and then revise them until the project and its intent is briefly and clearly explained.

1) My research will contribute to debates that consider…
2) My project will investigate…
3) Thus, my dissertation asks…

In light of question 3, it is useful to use to think of the particular descriptor that will be useful for outlining what your project seeks to achieve. For example, does your question ask **'how?'**, **'why?'**, **'where?'** or **'what?'** Using these words helps to frame your question.

Once you have read the next sub-section of the chapter, go back and look at your answers again. Do your statements reflect the research you intend to conduct and the type of knowledge you seek to contribute? Could you hone and improve your responses further?

Framing your question

Once you can express your research problem clearly and succinctly it is important to assess whether it is framed – or worded – as best it could be. Although your research question is normally just a line or two, it will be the foundational part of your dissertation. It will be the statement you refer to when conducting your research. It will be the basis of how you explain your research to participants (if your research involves working with people). It will be the benchmark against which the findings and contribution of your dissertation will be assessed by the examiner/marker who reads it. As such, it is vital that your question is framed in a way that accurately reflects the project you intend to conduct. This sounds like something of an obvious statement. However, a surprising number of questions do not convey the actual intent of student projects. The framing of a research question can be problematic when the following occur:

The question is too broad and does not reflect the specific, well-defined research area of the proposed project.

A student has been inspired by literatures concerning the nature/culture dualism (see Braun and Castree, 2005). They explore the wide array of literature in this area before encountering more specific reading on urban gardening and allotments (see Hitchings, 2003; Longhurst, 2006). Considering how they might add a new contribution to this work, they become interested in roof-top gardens – a novel feature of high-rise city blocks. The student is interested in how these gardens disturb ideas of a clear nature/culture divide, and the ways in which they might help to improve the aesthetics of the city. The student phrases their question as follows: 'My project considers the relationship between nature and culture in the UK.'

TASK 4.3

A good question should define the study accurately. Can you better rewrite this question for the student project outlined?

The question is long-winded and complex, so the meaning of the proposed project is lost.

Staying with the example outlined above, the student seeks to hone their research idea into a discrete research question, or statement. The student writes the following: 'My research aims to consider how gardens can challenge our assumptions about a long-held dualism between nature and culture that has persisted for some time in geographic thought, as well as to ask how these gardens (roof-top gardens in particular) can help in improving how city spaces look and feel for people who live there.'

TASK 4.4

A good question should be succinct and clear. Can you better rewrite this question for the student project outlined?

The question suggests the project seeks to find a definitive answer (yes or no, significant or insignificant, greater or smaller) when the project actually aims to investigate the complex, messy, contested social world that cannot be reduced to a single research 'result'.

Sticking with our current example of the roof-top garden project, the student frames their question in the following way: 'My research asks: Do roof-top gardens make for better looking cities?' The phrasing of this question suggests the student will, by the end of the dissertation, be able to definitively say 'yes, such gardens do improve city aesthetics' or 'no, such gardens do not'. Worded in this way, the marker of the dissertation would expect the student to have used methods suitable to make such a grand claim – perhaps drawing on statistical data (questionnaires or survey results) to make such a conclusive statement. If this is the kind of knowledge the student seeks to produce then the question is no doubt appropriate. However, if the student actually seeks to investigate the more subjective, complex and shifting engagements of city dwellers with these gardens, and to explore how people perceive them to impact the image of the city (using appropriate qualitative methods such as interviews or ethnography) then they will need to word the question quite differently.

TASK 4.5

A good question should reflect the kinds of knowledge we want to contribute. If the student is seeking to produce a qualitative study of nature/culture relations and engagements with roof-top gardens, can you better rewrite this question for the student project outlined?

Aside from using the appropriate wording in your research question, you might also need to think about how to frame your question in terms of its **presentation.** Research questions in human geography dissertations can be presented in different ways. Some students (and supervisors) prefer research questions to be very clearly phrased as a question – a problem that needs to be solved. For example:

> Is there a significant relationship between educational attainment and housing tenure in the west and east wards of Sheffield city centre?

Conversely, other students (and supervisors) prefer research questions to be presented as a statement or sentence that describes the topic. For example:

> My project seeks to examine whether there is a significant relationship between educational attainment and housing tenure. It will do this by looking specifically at the differences across the west and east wards of Sheffield city centre.

There isn't a 'right' or 'wrong' way to present your research question as long as it is there. Some students like to put the question in **bold** or *italics* so that it is clearly evident. Others might place the question in a text-box, or simply include it as part of a larger paragraph. This decision is yours and in part depends on the 'tradition' of your human geography dissertation. Indeed, often more quantitative or statistical studies will present the question clearly as a question; qualitative studies can tend to include the question as a statement or sentence. You can check with your dissertation supervisor how they advise you to format your question, but also consider what might work best in light of your independent project.

Writing aims and objectives

Finally, research questions tend to be accompanied by a set of aims and objectives. Aims and objectives can be thought of as sub-questions. They are more specific statements that help us to outline what we seek to achieve through our main research question. In other words, they are a set of goals that your project seeks to meet in responding to your central concern. Let's use an example. A student is conducting a project that considers the impact of the 2012 flood defence plan on the local village where he lives. The research question is as follows:

> How have the provisions of the 2012 Berkshire Flood Defence Plan impacted the resilience of the local village of Datchet against natural disasters?

In order to answer the question, the student has set the following aims and objectives that will allow him to respond to this research problem:

Aims and Objectives

1. To analyse the 2012 Berkshire Flood Defence Plan and ask how the provisions outlined help Datchet to better pre-empt and respond to flooding.
2. To speak to local planners to find out how the 2012 plan has impacted their activities in relation to flood defences and if any improvements could be made.
3. To speak to local residents of Datchet to find out how the plan has impacted their responses to flooding (if at all).
4. To consider how theories of resilience help us understand the impact of the plan for the village of Datchet.

Again, aims and objectives can be formatted separately in a list (as above) or might be presented in a paragraph that describes your intent. Simply consider which format helps you to explain the goals of your project most clearly. For example, the aims and objectives of the project could also be written like this:

> My project aims to analyse the 2012 Berkshire Flood Defence Plan and ask how the provisions outlined help the local village of Datchet to better pre-empt and respond to flooding. I further intend to speak to local planners to find out how the plan has impacted their activities in relation to flood defences and if any improvements could be made, and also to local residents to find out how it has impacted their approaches to flooding (if at all). Finally my project aims to reflect upon how theories of resilience can help us understand the impact of the plan for Datchet village.

As we can see from this example, aims and objectives, however they are written, are helpful in further **specifying your project**. They urge us to actually think about how we might conduct our project (in other words how we might operationalise our research in terms of what data we might collect, or people we may want to speak with). They also act as a neat reminder of what we are trying to achieve so our research stays focused as we actually 'do' our project (see Section II of this book). Like the research question, they are also a helpful measure for markers/examiners to consider. Your human geography dissertation should cohere from start to finish. Your research question should be the line of enquiry that your results, analysis and discussion actually 'answer' (and remarkably, sometimes questions and answers do not match in student projects). Aims and objectives therefore provide a clear statement of the details of your project that can be assessed when the research (and your discussion of it) is read, as well as keeping your project 'on track'.

Chapter Summary

- Developing a research question relies on reading. Once you have identified a broad area of interest, reading helps to narrow the topic down to something workable. Even if your idea is more defined, it enables you to contextualise ideas in broader debates to demonstrate your knowledge of the field of work in which the project sits. Reading ensures you take an academic and informed approach to your study.
- It is possible to search for relevant readings in several ways. Starting with your rough topic area (or even specific area of study) you can ask your dissertation supervisor, draw upon reading lists from lectures, use dictionaries, textbook entries and the reference lists of journals to find further readings, and draw on citation indexes and web-searches to find useful literature.

- Keep reading until you fully develop your question. Developing your question relies on engagement with the reading. Engagement with the literature allows you to ask if there are further avenues of study (that the author suggests or that you can see for yourself); if you might apply the ideas you read to a different case study, event, or group of participants; if the findings you have read about might have changed over time; or if there is a gap where work needs to be conducted.
- Once you have a firm idea, this needs to become a clearly articulated question or sentence that describes the purpose and intent of your project. Ensure your question actually reflects your project and that it accurately conveys the kind of knowledge you want to contribute to geographical scholarship. Aims and objectives can be helpful in further defining your question by setting a clear list or statement of the goals the project will meet.

Key readings

Flowerdew, R. (2005) 'Finding previous work on the topic', in R. Flowerdew and D. Martin (eds) *Methods In Human Geography: A Guide for Students Doing a Research Project* (second edition). London and New York: Routledge. pp. 48–56.

Hart, C. (1999) *Doing a Literature Review: Releasing the Social Science Imagination*. London: Sage (particularly Chapter 1).

Heley, M. and Heley, R. (2010) 'How to conduct a literature search', in N. Clifford, S. French and G. Valentine (eds) *Key Methods in Geography* (second edition). London: Sage. pp. 16–34.

Parsons, T. and Knight, P. (2005) *How to Do Your Dissertation in Geography and Related Disciplines* (second edition). London and New York: Routledge (particularly Chapter 4).

5

FINAL PREPARATIONS: IS YOUR PROJECT WORKABLE?

CHAPTER MAP

- Geographical explorations
- Practical considerations
- Assessing health and safety
- Making sure it's legal
- Paying attention to ethics

Geographical explorations

The Norwegian explorer Fridtjof Nansen set about an exceptional voyage in 1893. He was one of many figures involved in a race to reach the North Pole at the end of the century. At the time the Pole was the *terra incognita* (or 'unknown land') that had thus far eluded scientific discovery. As explorers and adventurers gradually mapped, charted, studied and recorded parts of the world that had previously been blanks on a map, the North Pole remained out of reach. Yet being the first to venture to the 'farthest point North' was an important achievement, the pinnacle perhaps, to define an 'age of exploration' (Nansen, 1897). Nansen's voyage was in many ways unique. He was inspired by the story of the *Jeannette,* a ship that had become stuck fast in ice and abandoned in 1879. The ship, ill-suited to the demands of the Arctic environment broke up amidst the ice, but not before having drifted with that ice for over two years. Nansen was intrigued by this movement. A ship that seems to venture North in accordance with a natural drift current led him to believe that the very same drift could be harnessed in a bid to be the first to reach the North Pole. Nansen developed what would become a defining project, of both success and failure. He set out to follow the voyage of the *Jeannette* – to build a ship, the *Fram,* which was more suited to the harsh environment – and set it to drift with the ice, recording, measuring and identifying these hitherto unexplored realms (Figure 5.1).

In order to undertake such a voyage Nansen had to contemplate the practicality and feasibility of his proposed research endeavour. For a start, this project was going to be costly. Nansen did not have the funds. Therefore, he set about working out whether he could raise the necessary cash to carry out the research. Nansen first put this case to the Norwegian Parliament and next to the Royal Geographical Society in London (Nansen, 1897: 41). He received funding from both bodies along with other public and private donors. However, Nansen had not anticipated the true cost of the undertaking and he had to support a portion of the voyage with his own monies (Nansen, 1897: 54).

Cost wasn't the only issue for Nansen. Perhaps the most important consideration was equipment. Nansen sought to ensure the safety and viability of his research with the very best piece of equipment of all – a ship, the *Fram* – which was suited to a long and arduous voyage in pack ice. Nansen also had to ensure he had a host of other equipment – sextants, compasses, maps, warm clothing, and food supplies – so he and his crew were ready for a voyage that could take many years to complete. He also had to ensure that he had a crew who were willing to participate in the research. This meant accessing and recruiting a skilled collective of explorers and scientists to assist in his voyage of discovery.

Figure 5.1 Preparing for fieldwork: Nansen's voyage on the *Fram* took several years of planning – negotiating issues of cost, access and the recruitment of suitable participants. (Source: Fridtjof Nansen (1897) *Farthest North,* Vol II. London: Constable & Co [Public domain], via Wikimedia Commons)

And Nansen's research was not without an ethical or moral dimension. Once the research had begun and the *Fram* was adrift, Nansen was increasingly concerned that he had misled his crew and his funders in Norway and at the Royal Geographical Society. In the 'race for the Pole' – with both personal and national pride at stake – the voyage was laboriously slow. At times it appeared the ship was moving no further north at all. He felt he had deceived those who had put faith in his plan, and backed up that faith financially.

Now it may seem, with the story of Nansen's famous *Fram* voyage, that we have strayed from the topic of *your* human geography dissertation. But as I have illustrated, however good our ideas may be, and however well situated they are in previous scholarship, we have to ask vital questions of practicality and manageability in order to assess whether our project will be workable. For Nansen, such preparation – in respect of funding, sourcing equipment, recruiting participants, planning the timing of research, meeting safety concerns, and reflecting ethically on his approach – was essential to the success of the research. Indeed, Nansen's work did prove the presence of a trans-polar current and his scientific works helped to develop oceanography as a discipline. Accordingly, the final, vital stage of designing a dissertation is asking the seemingly simple question: 'Can I actually do this study?'

This chapter covers the key factors you must consider when making this assessment. The chapter unfolds in the following order. We begin by examining the importance of access, travel, cost and time. We then turn to more specific practical concerns regarding health and safety, and the legality of your project. Finally, we discuss research ethics, examining what is meant by 'ethics' and the importance of conducting research transparently and sensitively.

Practical considerations

Many books on research practice tell us how to go about employing methods for data collection (and we will be covering this too in the next section). But few dedicate space to a discussion of the more practical considerations that go hand-in-hand with data collection. It might seem that you are jumping the gun by thinking of practical concerns before you have ironed out the methods you might use (more of which in Section II). Yet it is crucial at the design stage to think about the likely practicalities that will go hand-in-hand with the project you propose, to ensure such a project is 'doable'. Each dissertation is unique, so it would be impossible to provide a concrete list of practical concerns that you must contemplate. However, the considerations that follow are typical practical issues that are worth thinking through before you embark on your project.

Access

An obvious practical consideration to make when designing your research relates to access: to your research site, potential materials, or desired participants. Let's begin with the first. Geographical research, as we know, often involves going into one field or another. In other words, geography dissertations tend to have a distinct *geography* to them. In order to complete your research you may have to visit a particular place in order to conduct interviews, or gather questionnaire results. A useful starting point to project design is to ask if you can access your research site and if there are any special requirements that you might need to meet to do so. Let's take a couple of examples. Research that considers the relationship between architecture and consumption practices might rely on fieldwork in shopping centres. Certainly we all walk freely into shopping centres when we wish to buy something that we need. Yet shopping centres are not public spaces. They are semi, or quasi-public, in that they are open to the public but are typically privately owned. You will notice this if you look carefully the next time you visit such a space. Signs on entrances alert us to fixed opening times and a set of behaviours we must abide by (no skateboarding, no loitering, no smoking). Our behaviour in such spaces is often surveilled by private companies through CCTV and the presence of security guards. Thus, we would not be able to simply turn up and quiz passers-by with a project survey. Considering your research site in advance, then, and how you might access it, allows you to plan ahead. In this instance it might be advisable, for example, to contact the shopping centre manager and inform them of your work, asking if you might have permission to complete your research in that space. Having an information sheet (on university headed paper) that outlines your study can help in providing a clear overview of your study that can assist in gaining access and in recruiting participants. It is also part of good research practice (Task Box 5.1).

TASK 5.1

Writing an information sheet about your dissertation project is a useful task at the start of your project. Information sheets should articulate your research idea, precisely and concisely, to a public (non-expert) audience. They can help demonstrate an openness and transparency about your project goals. This helps to build trust. Using university headed paper is also beneficial for an information sheet as it legitimises your proposed project.

Write an information sheet (no more than one side of A4, the briefer the better) that outlines the focus of your study. It should state, clearly and succinctly:

- Who you are;
- Which institution you study at;
- Why you are conducting the study;
- What the study is about;
- How you can be contacted for further information.

Typically an information sheet will be presented with a consent form (which we cover later in the section on 'ethics').

Go online! Visit **https://study.sagepub.com/yourhumangeography** to see an example of an information sheet you might provide when gaining access or recruiting participants for your research. This can be used as a template for your own project.

However, access isn't simply about the research site itself. A project which relies on archive records must still consider access issues. First of all, you would need to work out which archive holds the material that will help you to answer your research question (see Chapter 7). Once you have identified your archive it is worth finding out if there are any particular requirements for accessing the records you need. Most archives are readily open to the public but often do require you to register for a reader's card (akin to a library card), in advance. The National Archives is one such example. This means you often have to take proof of ID with you. Knowing this in advance is helpful so that you don't waste research time by turning up without the appropriate documentation. Other archives (often smaller, or more specific collections) may require you to have a letter of recommendation from your department to support your application for access (the Lloyds Shipping Archive is a good example). Archives also have all manner of other rules relating to access. It is often the case that certain materials are prohibited (typically no pens are allowed, no erasers, and no external material that might be muddled with archive documentation). Once you have accessed the archive particular rules have to be met (these relating to the use of cameras, laptops and mobile phones). Considering access in advance helps you to ensure you can conduct your research without hindrance.

You also have to think about access to further materials you might need. If, for example, your project seeks to explore the role of policy in governing the behaviour of walkers in the countryside, you would need to ensure that the policies you need can be accessed. You might also want to check to see if any documentation exists that foregrounds the policy you are examining (minutes of meetings, public forums, and so on). The internet can be a fantastic resource for finding materials and it is useful to conduct a scoping exercise to check (1) that the necessary materials are available, (2) that they are not restricted under the Freedom of Information Act, and (3) that you can budget accordingly, should you need to purchase any documents (some policies are only available to buy, so you may need to consider the cost and whether a purchase is reasonable).

Finally, it is also worth considering your access to research participants if your project will rely on speaking with or working with people (individuals or groups). There can be all kinds of stumbling blocks to working with people. Access can often be tricky, even if it might seem more or less straightforward on paper. People can be unpredictable. They might not reply to your email. They might say they will meet and then not turn up. Social science research relies on both flexibility and persistence.

The good news is that people can also be incredibly helpful and interested in what you are studying, and therefore more than willing to assist in any way they can.

When designing your research it is worth jotting down who you might need to speak to, or work with, to address your research question. You can then assess the likelihood of being able to conduct the research or whether you need to consider an alternative plan. You can also think through various strategies or best approaches for making contact. For example, your research might rely on talking to someone 'elite' – for example, a governmental official. It is useful to work out in advance if you can realistically expect to access the person (or if there are other people who might also be able to provide the information you need). That's not to say you should not be ambitious. Often people are more than willing to help students with dissertation projects when asked. But it is necessary to balance ambition with realism. You are unlikely to get an interview with the Prime Minister, for example (should you think you need one to discuss national policy). However, you might consider contacting the relevant government department or your local MP.

Considering access also helps in thinking through the channels of communication that might ensure your access. If you are seeking to speak to software programmers about virtual space and cyber security, for example, an email might be a more than suitable way of getting in touch with the company or individual identified. In other cases a written and posted letter, or a phone call might be better. You need to think carefully about how to approach participants as first impressions count and can often open the door to the access you require. In addition, Gill Valentine (1997: 114) states that sometimes it is not possible (or even desirable) to access participants directly. We may have to negotiate access through a **gatekeeper**, or through a process of **snowballing** (see Information Box 5.1). Access is also dependent on building rapport and trust. It should not simply be expected after a brief email or conversation, or on the first time of asking. Ethnographic study, for example (that is, a period of research with a given community, see Chapter 7), is particularly dependent on building an honest and open working relationship with potential research subjects. This is because of the intensive and often intimate nature of work that involves the researcher integrating themselves into a new place or community.

INFORMATION BOX 5.1
GATEKEEPERS AND SNOWBALLING

There are two crucial ways in which we can gain access to potential research participants. The first is through a gatekeeper and the second is via snowballing.

Gatekeepers are those individuals who have the potential to unlock access to people you wish to talk to in order to answer your research questions. Often it is useful to gain access by using a gatekeeper rather than trying to contact the person you need directly. This is especially helpful when trying to access public figures, or if you have no contact details for the person you hope to access. For example, if you wished to talk to the CEO of an ecosystems services company to explore the role of businesses in supporting ecosystem provisions, you should probably contact the company PA first. Valentine (2005: 116) notes that using gatekeepers is often essential to professional research practice as a geographer. When contacting a gatekeeper you need to 'make it clear exactly what sort of information you want and who you would like to talk to' (Valentine, 2005: 116).

Snowballing refers to a process of 'using one contact to help you recruit another contact, who in turn can put you in touch with someone else' (Valentine, 2005: 117). That said, it first relies on gaining an appropriate contact (perhaps via a gatekeeper) who 'gets the ball rolling'. Once you have their trust, this becomes invaluable as you are 'recommended' to others who may be interested in partaking in the study. As such, it is important to build good relationships with participants at the start and show professionalism in order for them to put you in touch with others. However, some caution should be employed with snowballing. Like an actual snowball that grows in size, it might also move in unanticipated directions. You have to consider where a snowball of contacts may lead, and consider if the contacts gained are still relevant to the research being conducted.

TASK 5.2

Ensuring access might be straightforward for your project – especially if your starting point is someone you know or a place you are already familiar with. However, what if you want to access someone you don't know? What if you plan to research a place that you have no previous connection with? This task asks you to consider a set of people and places that could feature as part of a human geography dissertation. Ask yourself (1) how you might access the person or location and (2) what limits to access you might encounter and how you could overcome them.

People:

- Your local MP
- A member of the tourist board
- The CEO of a global company
- A celebrity
- A protest group

Places:

- A shopping centre
- A museum
- The London Underground transport network
- A planning meeting in your local area
- A youth community group

Return to this task again at the end of the chapter and consider your answers again. What health and safety hazards might you associate with researching these people and places? Are there any legal or ethical issues that might impact your access or approach to gaining access?

Travel

Closely related to the issue of access is that of travel. Assuming that your research takes you into the field means that you will likely have to travel to access the places or people you need to complete your study. That said, travel, for the sake of travel, does not guarantee a good dissertation. Your work does not have to be 'exotic' to be good. It might be the case that your project calls for overseas travel. But it may also be the case that it does not. It might be that your 'field' of study can be completed in the armchair, at home (see Aitken, 2005) or through virtual travel (Phillips and Johns, 2012: 10). The most important question to ask, is not 'is my location exciting enough?' but 'can I travel where I need to in order to complete my dissertation work?'

When considering travel arrangements it is worth drawing on any existing fieldwork/field class experience you have gained during your studies so far. Reflecting on how you conducted data collection activities, accessed locations and approached participants can help in your own independent research. That said, although you might usefully draw on past fieldwork experiences, your own dissertation fieldwork will likely be very different, especially if you are working independently. Fieldtrips at university are typically group affairs. They can be fun, and the group scenario provides a setting to 'let off steam' at the end of a hard day. It is important, if you are travelling for your dissertation, that you have someone to speak to at the end of the day. It might be a travelling companion, a friend at home, a family member, or your supervisor. In respect of the latter, check their availability and let them know when you are travelling, and ensure that you know what channels of communication are available in the destination you visit (and if you will have Wi-Fi, phone signal, and so on).

Finally, at the design stage of your dissertation it is crucial to consider any limitations to your proposed travel. If specific travel is necessary for your project, it is worth scoping out how long you need to travel for, the costs involved, and the health and safety implications (all covered in the sections to follow). This is as much a question for local travel as it is for long-haul destinations. Work in a local archive might still involve significant daily travel, coordination of bus timetables or negotiation of busy roads. For travel that does take you abroad it is certainly worth checking your government website for travel advice before you go. For example, in the UK there is comprehensive travel advice related to each global destination available at https://www.gov.uk/foreign-travel-advice. This helps in establishing:

- Whether travel to your given destination is safe and advisable;
- The entry requirements you require to a given state (passports, visas, and so on);
- The local laws and customs (which may be relevant depending on who you are and also your planned topic of study);
- Any health considerations you must make and the necessity of vaccinations;
- The currency you will need.

Cost

Talk of currency leads us neatly to a discussion of cost. In terms of monetary costs, it is worth completing a budget in advance to see if you can afford the study. Scoping out costs is useful in advance as you can consider (1) if you can afford to go, (2) if your budget is tight, whether you can make your trip more cost effective, (3) if you need to reconsider your project case study or location in regard to the cost. Costings should try to take into account the following:

- Travel (bus, train, plane, and also local travel and transfers, and any visas required);
- Accommodation (hotels, hostels, lodgings);
- Subsistence (money necessary to eat and drink during your trip);
- Insurance (suitable to cover anything of value you might take);
- Equipment (notepads, cameras, laptops, Dictaphones);
- Necessary research materials (printing of information sheets, consent forms and so on);
- The length of your trip (to account for cost of accommodation, food, etc.).

Go online! Visit **https://study.sagepub.com/yourhumangeography** for a blank copy of a budget planner. You can download this and fill it in to plan your costs, or print it off and complete if you prefer.

In respect of costings, there are several ways you may be able to save money in respect of any potential travel. The first is through equipment hire. Check your university facilities, as some institutions will have research equipment that can be loaned to students for research purposes (such as Dictaphones, video cameras and so on). If such a service is available (either in your department, or in the wider university) it is useful to make an enquiry as it could save you a significant sum. Secondly, many universities now offer travel bursaries to support student travel. Given the emphasis on fieldwork that is so central to geographical study, departments might offer small grants to support dissertation work. Check your institutional website or ask your supervisor to see if such funds exist. Whilst not all universities have such provisions, all students are able to consider the funds available from external sources. The Royal Geographical Society, for example, offers a range of awards for undergraduate and postgraduate students to support dissertation work and field travel (see their website at http://www.rgs.org). Although incredibly competitive, if you have a very well designed and planned project that requires funds for travel it can be useful to discuss the opportunity with your dissertation tutor and make an application. Be aware of any deadlines for travel funds and bursaries so you don't miss potential opportunities.

Time

You will be well aware from the very start of your dissertation project that time is a crucial consideration. There are many long- and short-term considerations you must make when designing a schedule of research. In the longer term you must consider if your project is do-able within the entire time frame available. This means filling in the blanks of a calendar with the differing tasks necessary for your project (reading, planning, doing research, analysing, writing up, revising and editing, and so on). As noted in Chapter 1, mapping out a realistic timetable at the start can be helpful in identifying the feasibility of your study. It is always advisable to build a degree of flexibility into your timetable at the start in case you need to change track with your project. It is also useful to conduct a **pilot study** at the start of your research to test the waters and find out if your potential project is workable (Phillips and Johns, 2012: 53–4). A pilot might include scoping out your study area; assessing how long it might take you to complete your interviews, or exploring the number of files in an archive to gain a measure of the volume of material you will have to look at. Typically a pilot also includes trying out your questionnaire or interview questions to see if they are suitable for their intended purpose.

It is also beneficial to consider the time of year you aim to complete your research. For example, a student decided to complete her dissertation on the topic of sporting culture asking how cultural identities are constructed and performed on the football pitch. The student was hoping to carry out her research during the summer months, collecting her data in the period ahead of her final year at university. However, the student omitted a major consideration from her plan. During the summer – when she was scheduled to complete her data collection – the football season had ended. Accordingly, where your research has a particular temporal component you may have to think more carefully about when to conduct the research and if it is even possible in the time available.

Time of day can also feature as important in designing your dissertation. Another student was interested in urban space and the night-time economy. For his planned research, he sought to examine the rules, regulations and practices that governed those using city spaces to ensure public safety and security. The project posed some particularly tricky questions. Much of the dissertation work required him to work independently. But would he be safe at night – particularly in city centre hotspots that were known for 'trouble'? Research that considers spaces of dark and light (see Edensor, 2013) and night-time spaces alerts us to the importance of time to our research activities. Of course we can do research at night but it might require us to take extra precautions. Accordingly, identifying this in advance helps us plan and mitigate potential risks. Emma Spence discusses how she negotiated issues of time, and also access, travel and the costs associated with research in the Graduate Guidance box below.

CONSIDERATIONS FOR WORKING IN A FOREIGN FIELD
EMMA SPENCE

As an undergraduate student my summer job was as a crew member on board various luxury superyachts in the Mediterranean. When pondering initial dissertation ideas I was inspired by Philip Crang's (1994) paper on conducting ethnographic research in his place of work. Encouraged by the possibility of combining casual employment with geographical research, I knew right away that I wanted to study the luxury superyachting industry. I discussed my research aims and objectives with the crew before arriving to the yacht, and openly discussed my observations once on board. This overt approach with the crew was essential in order to establish trust and in ensuring participation (see Spence, 2014; and also Chapter 7). Working full time, being continuously mobile, and often out of reach of shore led to many practical challenges that ultimately shaped my research along the way.

Conducting ethnographic research in my place of work meant that significant costs, typical of conducting research in a foreign field (including accommodation, travel, and insurance), were taken care of. Travelling with a British passport enabled the free movement between nation states in Europe without visas. That said, before leaving I had to ensure that I had plenty of time left on my passport to avoid any difficulties with various port authorities.

Securing employment on board in order to conduct my research relied upon on having up-to-date sea survival, fire fighting and medical certificates. I already held a valid sea survival certificate and so avoided the significant upfront costs that would have otherwise hampered my research. The yacht is a highly regulated working environment and in addition to sea survival training the crew had weekly emergency drills. Being away from the rapid response of terrestrial emergency services made conducting research at sea at bit more precarious. My risk assessment completed months beforehand was testament to this.

On board we were fortunate enough to have good internet access, so I was able to keep in regular contact with friends and family via email and Skype. In the field I purchased a pay-as-you-go SIM card to use in an old mobile so that I had multiple low-cost ways of keeping in contact at sea and ashore. Email became invaluable for keeping an open line of communication with my dissertation supervisor. Despite being hundreds of miles away and bobbing around at sea, regular contact enabled me to keep my research on track.

Working full or part time and conducting research at the same time is not the easiest thing to do. I had to be disciplined with my free time and ensure that I was making good progress with my dissertation. Having said that, I took full advantage of being in amazing locations, with a fantastic group of people from all over the world. Research can and should be hard work, but with the right project and appropriate pre-planning it can and should be a lot of fun too.

Assessing health and safety

Researching at night brings to the fore an essential practical consideration for all dissertation projects: health and safety. For human geographers, health and safety can often seem unimportant. What, after all, might be the risks of going to conduct research in an archive? Or of conducting a focus group with ethical shoppers? Bullard (2010), Flowerdew and Martin (2005), Phillips and Johns (2012) and Valentine (2005) (all incredibly experienced geographers) urge students to take health and safety seriously.

At the design stage, we should assess the practicality of our project in view of the health and safety implications it might entail. Institutions have a duty of care to you as a student. Any fieldwork associated with university business must comply with specific regulations. That said, ultimately *you* are responsible for ensuring you complete the necessary forms carefully and thoughtfully, and undergo any training that is required so that you are adequately prepared for your research (see the Graduate Guidance box above). As Joanne Bullard states (2010: 49), health and safety involves 'the practical steps that geographers can take to lessen the chances of an incident or accident causing harm to themselves and others during fieldwork'. Check your institutional guidelines for health and safety procedures, but typically a risk assessment is made up from the elements outlined in Information Box 5.2:

TASK 5.3

ASSESSING HEALTH AND SAFETY

1. Identify hazards: Jot down a note of the things that could prove hazardous during your research. Try to think outside of the box to ensure you are as thorough as possible.
2. Consider who could be harmed: Think about who (or what) could be harmed by the hazards you have identified. Is it you? It is your participants? Is it the environment you are working in?
3. Assess the risks: This is where you must question the likelihood of the hazards occurring and the 'severity' of these (at my institution students have to give a number to their hazards, ranking them from 1 (very low risk) to 5 (very high risk). Although this system might not exist at your own university it can be a helpful way to make a judgement call about how risky you think certain elements of your research might be.
4. Suggest how they could be avoided or minimised: Once you have identified and ranked potential hazards, think carefully about how the risks they pose can be avoided and minimised. What measures can you take to protect yourself, your participants and your research site?
5. Make a written record: Professional research practice dictates that we should keep a written record of this assessment and the precautions we seek to take. This is normally a formal requirement.
6. Review your assessment at regular intervals: Research is not a static process. Places, situations, and the people we work with are constantly changing. Hence, hazards and associated risks can change too. Reviewing your assessment therefore, as your work progresses is vital to professional (and safe) research practice.

(Developed from Bullard 2010: 50; Phillips and Johns, 2012: 63)

A central consideration in many risk assessments is the matter of lone working (Bullard, 2010: 52). Producing an independent piece of work typically means you will end up researching on your own. Thus, you must take the implications of lone working seriously – especially as a human geographer – where your research may involve working with people who could be unknown to you. Notably, where research requires us to meet with strangers for an interview or focus group, you must conduct research in a place where you will be safe. As Valentine notes, social science researchers should remember issues of safety: 'take your own ID, ask for theirs and meet in public places' (2005: 118). When working with people, the following measures are essential:

- Planning to meet your participants in a public place;
- Always letting someone know where you are;
- Taking a phone with you (and making sure the battery is charged!);
- Looking after yourself emotionally.

In respect of the last point, it is worth knowing that it is not only participants who have the right to withdraw from research (see the section on ethics to follow). We can truncate research too. It could be that something we have heard in an interview or as part of ethnographic research has bothered us greatly. It could be that we have been made aware of something troubling or even illegal. In the next sections we consider the practicalities associated with conducting research legally and ethically.

Making sure it's legal

Over the past two decades geography has experienced a 'legal' turn (see Blomley and Delaney, 2001). Scholars have become particularly interested in the operation of law and how legal frameworks shape spaces and places and the actions of those using them. Whilst this work is undeniably important, there is a notable absence of critical discussion of the role of law in our research practice as geographers. However, the legality of our research endeavours becomes all the more important in an age of 'action-orientated' research that is more activist in nature (see Chapter 8). What is interesting about such research is that in examining the law – and, on occasions, transgressions of it, or exploring the groups that may be positioned on the edges of law – researchers can be thrust into legally complex situations themselves. Research is not always neat and tidy, or indeed safe. If we are to attend to the most pressing social, cultural, political, economic or environmental concerns impacting our world, we may have to go outside of our comfort zone. Indeed, some of the most exciting geographical research in recent years has been located in muddy legal waters. However, in spite of such agendas, we must still consider whether our research is possible on legal grounds as it remains that we have a responsibility to ourselves, our participants, our supervisor and to our institution to work within the law.

Projects that involve illegal actions should certainly be avoided at the outset. Your university will not be able to offer protection against activities that break statute laws. However, you should also consider at the design stage if your project could involve illegal activity (trespass, illegal protest, vandalism, and so on). If you suspect there could be a risk of illegal activity you can plan your work in a different way to ensure no breach of the law should occur (for example, you could turn a contemporary project into a historical one using secondary data, circumventing legal conflicts that could arise).

However, you may find that your research inadvertently brings you in touch with illegal activities during the data collection stage. For example, if a community group, with whom you have developed a long-term working relationship, asks you to help them with something legally tenuous it can place you in a difficult situation. On the one hand you might want to help an individual or group you have come to know. On the other hand you may be frightened about the ramifications of involvement (or indeed by refusing involvement). You may also see or hear something that troubles you. As a researcher you have a duty of care to your participants. This includes an obligation to protect the confidentiality of participants and to take their comments 'off the record' if they ask. But what happens if you find out that a participant has committed a crime? Where does your responsibility lie then? Like medical doctors, we still have an obligation to break a clause of confidentiality if there is a greater harm likely in not doing so. Such events are unlikely but they are possible. In these instances it is best to consult with your supervisor or a suitable authority from whom you can gain advice and support. Such events also raise ethical, not just legal, issues, and it is to this the chapter now turns.

Go online! Visit **https://study.sagepub.com/yourhumangeography** to access a trio of Sage articles from the journal *Qualitative Research* that offer insights into practical concerns that shape research practice in light of health and safety, legality, and the importance of conducting pilot studies.

Paying attention to ethics

The case of Suman

In a recent paper in the radical journal *Antipode,* academics Richa Dhanju and Kathleen O'Reilly reflected on a particularly difficult and upsetting aspect of their work as development geographers located in Rajasthan (2013). As academics, and *outsiders* to the community they were working in, they grappled with how best to attend to the fact that a child, the daughter of a woman they are interviewing, appeared to be severely malnourished. They were concerned about the well-being of the child. Their enquiries as to the health of the girl were rebuffed by the parents, and their offers to help in taking the girl to the hospital were rejected by the wider community. A neighbour cautioned the pair 'against … intervention' (Dhanju and O'Reilly, 2013: 514). They left. Yet their concern, and the moral obligation they felt, drove the pair to report the girl's ill health to a Child Development officer, who assured them that he would help. This was a decision to help the child – above the wishes of the family and the community, their research participants – to whom they owed transparency and honesty. However, there was a breakdown in communication. An officer had not visited the girl. This was in spite of the researchers chasing up the incident, time and again. And in the end, it was too late anyway. The child had died.

Harrowing as that incident was, it alerts us to the ethical dilemmas faced by human geographers. Dhanju and O'Reilly had to profoundly question their responsibilities. Certainly they had a responsibility to be respectful to the customs and traditions of the neighbourhood in Bukhasar village where they were working (2013: 513). Certainly they had a responsibility to follow the wishes of the participants who had granted them access to this remote place. But what if a situation arises – like that described by Dhanju and O'Reilly – where we find we can't respect customs or traditions, or where we are forced to break the promises that we have guaranteed? How can we make such a decision and whom do we harm in the process? Is the harm we might permit less than the harm we are trying to prevent? These are all ethical questions and such questions are essential to the practice of geographers, particularly human geographers. This example might seem a million miles away from any kind of ethical dilemma that you might face as a student. But, in fact, ethical issues present themselves frequently during research. You are unlikely to complete your own dissertation without dealing with one ethical quandary or another.

Being an ethical geographer

Before we can know how to be ethical, it is useful to think about what exactly ethics are. For this, it is useful to venture outside of geography and seek out a basic definition. The Oxford English Dictionary (OED) defines ethics as 'the moral principles that govern a person's behaviour [in] conducting an activity' (OED, 2013). As noted in Chapter 1, many of us become geographers 'to make the world a better place' – to do the 'right' thing (Hay, 2010: 36). Yet this moral rationale can jar with the process of research, because research is an enterprise that allows us the privilege to investigate something that is of interest or fascination to us. As is sometimes said – research is 'really quite an arrogant enterprise' (Agar, 1980: 1). So how do we begin to behave ethically as geographers?

Engaging with ethics is important at every stage of the research process; as you design your project, do it and deliver it. Let's consider in more detail how we attend to ethics in each of these stages:

Designing

When we design our research, we should anticipate the likely consequences of our work on the people and places central to our study. What might the impacts be? Is it likely, for example, in a project about urban crime, that we might upset participants in a discussion about this sensitive topic? When we plan we should identify what harm there might be and how we might minimise and mitigate against such impacts. Most universities will require you to complete an **ethics assessment** before you start your research. If there are any particular issues, your project may have to be considered by an ethics panel.

Doing

When we conduct our research, we must act ethically. This means gaining **informed consent** to talk to the people we want to speak to, or to enter the communities of those we want to study. Consent is more than simply permission. It is an agreement to participate based on the researcher setting out a framework that will protect the participant, and ensure openness and honesty during the research process (see Information Box 5.2 below for more details).

Go online! Visit **https://study.sagepub.com/yourhumangeography** to see an example of a form you might use to gain written consent. This can be used as a template for your own project but must be adapted to your own institutional guidelines.

Delivering

When we write up our research we have to ensure we represent our findings fairly. For example, we cannot invent or falsify data (simply because we didn't find out what we wanted to). We should not 'cherry-pick' our data to make a particular argument or take data out of context to support our claims (see Chapters 10 and 11). We should also consider the ramifications of including data if it makes our participants identifiable when we have promised anonymity. We must further question how we use data that might express sexist, racist or bigoted views.

Information Box 5.2 outlines the core ethical principles that guide researchers, developed from the Economic and Social Research Council (ESRC) (www.esrc.ac.uk) which is the major government funder of social science research in the UK.

The ethics framework set out by the ESRC (and other bodies) often feeds into university-level ethics guidance. It is this specific guidance that we must consult at the design stage of our dissertation to ensure our research is planned with these 'best practice' measures in mind. Make sure you look up guidelines specific to your institution as you plan your research.

INFORMATION BOX 5.2
ETHICAL GUIDANCE FROM THE ESRC

- Research should aim to maximise benefits for individuals and society and minimise risk and harm. This means participants should not be adversely affected by partaking in the research (physically, socially or psychologically).
- The rights and dignity of individuals and groups should be respected. This means protecting the rights and privacy of participants and ensuring them confidentiality and anonymity. Participants should be informed as to how anonymity will be ensured, and to what degree, or if this will not be possible. Research with vulnerable groups should be carefully considered.
- Participation should be voluntary and appropriately informed (i.e. with an information sheet). Consent should be obtained and participants should have the right to withdraw from research at any time (without needing to provide a reason).
- Research should be conducted with integrity and transparency. Participants should be asked whether research can be recorded using a Dictaphone (or other recording device) and be able to refuse if they prefer data-gathering devices to not be employed. Participants should be informed as to how the research will be written up and disseminated (and, increasingly, given the opportunity to review any work ahead of its publication or circulation). Researchers have a duty to communicate findings accurately and truthfully.
- Lines of responsibility and accountability should be clearly defined. This means that it should be clear to participants (i.e. in written form on an information sheet) who is responsible for the research you are undertaking and how they might contact that person if a query rises.
- Independence of research should be maintained and where conflicts of interest cannot be avoided they should be made explicit.

(Developed from 'Our policy and guidelines for good research conduct' (ESRC, 2016))

Go online! Visit **https://study.sagepub.com/yourhumangeography** to find a web link to the guidelines for ethics set out by the ESRC. You can also find other helpful web links here, relevant to ensuring your project is workable. These include links to UK government travel advice, Royal Geographical Society student funding opportunities, and health and safety advice when planning fieldwork.

Ethical dilemmas

Sometimes our proposed research can present us with dilemmas that confuse our moral principles (as the example of legal quandaries demonstrated earlier in the chapter). In such cases there are no set recipes 'for resolving ethical choices or dilemmas'; instead you must 'make such choices on the basis of principles and values, and the (often conflicting) interests of those involved' (Aberystwyth University Ethics Guidance, 2014). Go back to Task 5.2. How might the ethical guidelines in Information Box 5.2 above inform how you plan to conduct research with the people and places noted? What ethical provisions should you employ? Are there any ethical dilemmas you might face in respect of the people and places presented?

Ethical considerations are essential to all research projects, but ethics are never easy to assess, and nor can they be confined to one part of the dissertation alone. Whilst we have primarily considered the ethical dimensions you must consider at the start of the dissertation, to ensure your project is feasible, ethics – much like health and safety – should be reviewed throughout your research (especially if your project changes) and whenever any particular ethical dilemmas arise. Indeed, working ethically is just as vital in the process of doing your research as in its initial design. And it is the task of doing your dissertation which we turn to next.

Chapter Summary

- Research design involves assessing a host of very practical questions to ensure a project is practical and manageable. Considerations involve thinking seriously about whether you can access your research site, materials and potential participants. It also involves thinking about where you might need to travel (and when), the cost of your research and if you will have the time to complete your intended study.
- It is very important to consider the health and safety hazards associated with a project and to assess the risks that these pose. Doing so helps to mitigate harm to you and your participants. Universities will require you to complete a formal health and safety form.
- Research that involves illegal activities should be avoided from the outset. Hearing or seeing something illegal during your research, or being asked to engage with something illegal, requires you to act in a responsible manner. Seeking advice from your supervisor or appropriate authority is essential should you encounter such a situation.
- Ethics refers to the morals or principles that govern a particular activity. During all stages of our research we should abide with the ethical guidelines of our institution that inform us of best practice. We should also ensure we gain ethical approval for our project prior to commencing research.

Key readings

Bullard, J. (2010) 'Health and safety in the field', in N. Clifford, S. French and G. Valentine (eds) *Key Methods in Geography* (second edition). London: Sage. pp. 49–58.

Hay, I. (2010) 'Ethical practice in geographic research', in N. Clifford, S. French and G. Valentine (eds) *Key Methods in Geography* (second edition). London: Sage. pp. 35–48.

Phillips, R. and Johns, J. (2012) *Fieldwork for Human Geography*. London: Sage (particularly Chapters 2, 3 and 4).

SECTION II

DOING YOUR HUMAN GEOGRAPHY DISSERTATION

6

DOING REFLEXIVE RESEARCH: SITUATING YOUR DISSERTATION

CHAPTER MAP

- First-person politics
- The 'god trick' (and other myths)
- Situating knowledges
- Placing yourself in your research

First-person politics

Have you ever wondered why so many human geographers write in the first person? The very simple reason is that most human geographers believe that they are a part of the research they are writing about. It makes sense, then, to communicate research accordingly. Indeed, in my own work I have always used personal pronouns to explain myself (and this book offers no exception). As Atkinson and Shakespeare note (1993: 4), 'the self, the "I" is part of writing and research, and interacts with ideas and people'. In other words, the 'self', the 'I', is not separate from the work we do as geographers. So how do you begin 'doing' your dissertation research? A good starting point is acknowledging that *you* are part of the process.

This seems like an obvious point. After all, this book has stressed several times already that the dissertation is an independent project, one of which you, yourself, have ownership. To say that you will be crucial to the collection and interpretation of data seems redundant. Certainly it will be you who interviews participants, administers surveys and analyses texts. Yet as this chapter hopes to demonstrate, taking seriously your own role in doing research is vital to 'critically examining power relations and politics in the research process, and researcher accountability in data collection and interpretation' (Sultana, 2007: 376). Indeed, doing research is not an apolitical, neutral process. Doing research is to engage in a complex journey of gathering data and making sense of it, with which you, yourself, are intricately entwined.

The geographer Jesse Heley (2011) illustrates this point with aplomb. Whilst researching class, culture and rural life in Eamesworth, Heley found that as a local in the village he was studying – and at least in some ways familiar with the character of country living – his research was intimately influenced by this connection. Jesse the 'village boy' was not separated from Jesse 'the researcher'. As Heley argues, this connection was actually beneficial to the research in permitting him unrivalled access to participants, made possible through his geographical knowledge of the area, his credentials as a 'local', and ultimately his position as a barman in the local pub, 'The Six Tuns'. But whilst his position in the community made certain research avenues possible, this connection also shaped the generation of data. A different researcher – someone without the local lingo and knowledge – would have undoubtedly gathered different data. And although Heley's research focused on a particular aspect of society that he himself was not acquainted with (the so-called 'new squirearchy' of 'plastic farmers' and 'wannabee Hooray Henries' who embrace country life with an 'appetite for all things "gentry", from country houses to waxed jackets and shooting parties' (Heley, 2011: 224–5)) – his status as proximate, or close to those he worked with, allowed him to quickly build a 'strong rapport' on the basis of a shared belonging to the village (2011: 226).

Indeed, research and the researcher who conducts it are never separate. Whilst the research we read may seem authoritative, is has still been 'authored' (Feminist Pedagogy Working Group, 2002: 135). As Paul Cloke and colleagues put it:

> In the practice of human geography, research is based upon the 'construction' of knowledge, found, created and enacted by 'us' as people living and working within specific economic, political, social and cultural contexts. (Cloke et al., 2004: 4)

Often this comes as quite a radical realisation. As the Feminist Pedagogy Working Group note:

> For many of us, our experiences in learning have led us to believe that the contents of the books we read are true, that the authors are knowledgeable, and that somehow together they constitute authority. (2002: 135)

Recognising that the research you read and the research you will be conducting isn't definitive but is partial, biased, and shaped by you is a powerful tool in conducting more transparent, honest, and therefore respected research. But how do you go about doing what is called 'reflexive' research?

This chapter is concerned with a set of fundamental skills that human geographers must often develop when doing research – skills in positionality, reflexivity and situating knowledge. This chapter is deliberately placed at the start of the second section of this book as such considerations are now regarded as crucial to 'doing' good research. This chapter first seeks to understand how and why geographers have come to care about their own role (and that of others) in doing geographical research, introducing the 'god trick' and the myth of objective, all-seeing knowledge. It then outlines a key tool or 'technology' (Rose, 1997: 308) for doing research more critically – reflexivity – explaining how this process helps you take your position (and that of others) seriously in order to better situate the knowledge you produce. The chapter ends by exploring some of the more practical challenges of placing yourself into the research process, examining how you might actually go about 'doing' reflexive research.

The 'god trick' (and other myths)

Jesse Heley is not alone in contemplating how he had personally shaped his research in the process of conducting it. Whilst he interrogated his credentials as a 'village boy', Farhana Sultana similarly quizzed her position as a local, female researcher whilst completing a project concerned with resource management and access to clean water in Bangladesh (2007). It might be easy to assume that a consideration of who you are will play little to no role in a study of arsenic levels in groundwater and the implications of this for a rural community who are subject to such contamination (see Sultana, 2007: 377). Yet as Sultana demonstrates, the data she collected (or failed to collect) about the drinking water crisis was not separated from her own position in conducting the research. Yes, she was from Bangladesh. But in doing the research she 'was by no means returning "home"' (2007: 377). Sultana grew up in Dhaka, the capital city. This was miles from the small villages that were the focal point of her study. She was acutely aware that her appearance (short hair, trainers, watch) and her less visual attributes (class and education) set her apart from her participants. She notes how people were willing to engage with her in the research process but her role in the community was certainly as a 'guest' and she had to constantly negotiate her position as a researcher to 'build rapport' with her participants and allow her research to proceed (2007: 379–80). Ultimately Sultana discusses this complex process because in doing so it becomes possible to 'reflect on how one is inserted in grids of power relations and how that influences methods, interpretations, and knowledge production' (2007: 37).

As geographers we have a responsibility to consider how the research we 'do' shapes the knowledge we create. Doing 'reflexive' research ensures that your project takes into account the factors that shape it. This helps to produce a more accountable, transparent form of research that is comprehensive and convincing because it takes into account its limitations, its bias and the politics that shapes it (rather than claiming to be objective, indisputable fact). Such an approach dispels what Donna Haraway calls 'the god-trick' (1988).

Knowledge, as we know, is produced through research practice. And knowledge is powerful. Often it can be so powerful that we accept it without question. It is at this moment that knowledge becomes detached from the research itself and the person who created it. It is at this point that knowledge appears to be established by some all-seeing, god-like entity. And this makes such knowledge influential. It makes it appear as undisputed fact. Let's take an example and consider what we know about gender. Traditionally women are associated with a set of characteristics and behaviours (just as men are as well). Women, historically, have been constructed as weak, irrational and emotional, as well as being 'the keepers of family and morality' (Mitchell, 2000: 208). These characteristics have shaped the geographies of women's lives, often relegating them to 'their place' in the private sphere of the home (Cresswell, 1996: 227). Yet this characterisation (and location) of women is not 'god given' as the 'god trick' of impartial knowledge would have us believe. This knowledge about women is a social construct and a construct that has been carefully and purposefully forged over time, by people, for a purpose (Sayer, 1997). It has been a method of creating hegemonic patriarchy (or the dominance of men in society). It is not some abstract, objective fact that has emerged from fresh air. Now we might think that such ideas about women are consigned to the past – but knowledge about gender, and also race, sexuality, religion, age (the list could go on and on) create real world discriminations that persist to this day (see Cresswell, 2013: 146).

Thus, critically examining what we know is a vital thing for us to do. It is our job to quiz and question the world (see Chapter 4). We should not accept we know everything, but recognise there might be secrets we can uncover, phenomena we can explain. As the Feminist Pedagogy Working Group note (2002: 135), it is often easy to 'give way to dominant ideas', to get 'caught up in routine complacency, failing to question authors and authority'. Yet as Tim Cresswell has noted (in an effort to expose the power of ideas) we should be aware that things we take to be natural, rarely are so (1996: 18). As he notes, ideas often become distanced from the people who have forged them. Instead, they appear as common sense. The more common sense or 'naturalised' an idea becomes, the more difficult it is to see it as anything other than the 'truth' (Cresswell, 1996: 18). Yet it is our job as geographers to interrogate what we know and recognise how knowledge is always formed by someone, somewhere, for some reason.

Feminist scholars were among the first to challenge the vision of scientific objectivity, arguing that the construction of knowledge can never be impartial or unbiased because it is always exactly that – a construction (see Haraway, 1988; Harding, 1986). Feminists have argued that no knowledge is produced in an objective, value-free way. Indeed, feminists have argued that the knowledge we so often take as 'fact' is still a construct – a white, heterosexual, masculine construct (as typically white, heterosexual men have been the creators of knowledge) (Haraway, 1988: 579). As such, the idea that knowledge can be anything other than a partial product of the person doing the research is exposed as a myth. This can be a difficult idea to process (especially when we are so used to taking the knowledge we encounter at face value). But the point, more simply put, is that we should be alert to the **origins** of any given knowledge. We should be critical of the ways power creates certain types of knowledge. We should be open to alternative types of knowledge, produced by a range of different voices (often those that are less well heard).

Geographers such as Kim England (1994), Linda McDowell (1999), Gillian Rose (1997), Pamela Moss (2002) and Farhana Sultana (2007) have contended, for example, that women can produce radically different knowledge about the world because of their place and position. In other words, it is unhelpful to think about research as producing one infallible truth. Rather, the place and position of the researcher (as a woman, a man, as black, white, straight, gay, young, old, able or disabled) forms part of the knowledge that is produced – making that knowledge partial to the person who has produced it. Accordingly, when we conduct our own research, it makes sense that the resultant knowledge is created by us too (and is shaped by all the traits we might bring to it). Acknowledging this creates, as Donna Haraway writes, 'a more adequate, richer, better account of [the] world' (1988: 579).

Now this is all relevant to actually doing research because it alerts us to the ways in which **who we are**, shapes **what we do**. Part of doing geographical research requires you to think about how exactly your research is being shaped and informed. What parts of your identity or character might be shaping the research you conduct, the questions you ask in interviews and the observations you make in your research diary (Rose, 1997: 308)? As Jesse Heley's example demonstrates, his role as a local village boy mattered to the data he collected. In being 'local' he had the trust of his participants and this produced more 'privileged' access to those he wanted to work with (Heley, 2011: 227). His status shaped the process of doing research and ultimately the findings that resulted. Heley's account of the 'new squirearchy' of 'lords of the manor' and the 'country set' is a credible one because he did not pretend his research was without bias. In acknowledging his position and the partiality it brought to the research, Heley was able to write a more convincing, substantive and rigorous account of the rural classes in Eamesworth. The next section considers what we call reflexive research in greater detail.

Situating knowledges

In 1997, Gillian Rose set out to question how we might better 'do' our research as geographers. The result was her landmark paper 'Situating knowledges: Positionality, reflexivities and other tactics' (published in *Progress in Human Geography*). In this paper Rose critically discusses the move towards doing 'reflexive research'. The paper outlines what is meant by positionality, reflexivity and situating knowledge and raises some astute observations concerning the problems of trying to embrace this approach in practice. Rose's important paper is a must-read for any human geographer. The paper is available to read via the

Companion Website (see the Go online! link to follow). I'd recommend reading this paper. Once you've done so, consider the questions set out in Task 6.1.

Go online! Visit **https://study.sagepub.com/yourhumangeography** to access Gillian Rose's 1997 paper 'Situating knowledges: Positionality, reflexivities and other tactics' *Progress in Human Geography* 21 (3) 305-20. You can also access two other Sage journal articles online, which discuss the challenges and practicalities of engaging with reflexive research practice.

TASK 6.1

Read Gillian Rose's pivotal article in *Progress in Human Geography.* This paper is incredibly engaging but also very challenging. Stick with it as it will greatly assist your understanding of reflexive research practice. To help you grapple with the ideas presented in the paper, make notes on the following questions:

1) What is positionality?
2) What is reflexivity?
3) What does it mean to 'situate knowledge'?
4) What is the sense of 'failure' that Rose describes with this process?
5) What are the challenges of actually doing reflexive research?
6) How does Rose suggest we can work through those challenges?

Reflecting on Rose

What can be learned from Gillian Rose's pivotal paper? Reflecting on Rose's arguments helps us to grapple with what it is to consider positionality, to embrace reflexivity and to situate knowledge. Let's take each of these in turn.

Positionality

This refers to our social position or place. For example, we may find ourselves in a higher or more privileged social position than those we research (or it may be that the reverse is true). To consider our positionality we must ask how 'facets of self' – any characteristics that shape our position, such as 'race, nationality, age, gender, social and economic status' – might influence the research we conduct (Rose, 1997: 309). Rose notes that taking our position into account is vital. If we accept that knowledge is created from a particular position (a sexed, raced, gendered, aged position) then knowledge can no longer claim to be universal (to follow Haraway, 1988). Positioned knowledge becomes partial knowledge.

USEFUL TIP

When you do your human geography dissertation it is necessary to consider the position you hold, or the place you take in your research.

Reflexivity

This refers to the method we enact for 'avoiding the false neutrality and universality of so much academic knowledge' (Rose, 1997: 306). Reflexivity is a 'technology' or 'tool' of research, which encourages a critical reflection of our positionality and how 'facets of self' shape our research (in view of the methods and approaches we adopt for collecting data, and how, in turn, we analyse and interpret findings). Michael Lynch outlines several 'types' of reflexive practice but notes in short that as a method it is a process of 'self-reflection', 'self-consciousness' and 'self-criticism' that helps to 'root out' (Lynch, 2000: 34) and identify the bias that shapes our geographical knowledge.

USEFUL TIP

When you do your human geography dissertation employ reflexivity to identify and critically reflect upon how you, yourself, influence the research you conduct.

Situating knowledge

This refers to the production of knowledge that is situated within a discussion of its limitations and biases. Situated knowledge is knowledge that has been subject to reflexive consideration (see above). It is knowledge that has taken into account the politics of positionality. In this way, situating knowledge is an outcome of reflexivity. Situating knowledge is the *placing* of knowledge to subsequently produce findings that are more transparent and representative. As Rose notes (1997: 306), '"situating" is a crucial goal for all critical geographies' because it produces research that is credible on account of acknowledging the very factors that have shaped it.

USEFUL TIP

When you do your human geography dissertation situate your knowledge so as to add rigour by acknowledging its partiality and bias.

Yet Rose also raises further points for us to take into consideration. The first relates to how we practise reflexivity. Doing reflexive research is not purely a process of looking inwards to contemplate who we are and how our inner attributes and character shape the work we conduct. Reflexivity is also a process of looking outwards in order to place ourselves in the world and to understand our situation in it, since this too shapes the data we collect. This is what Moss calls the 'double reflexive gaze' (1995).

The second point that Rose urges us to think about is how this consideration of ourselves (looking inwards or outwards) is never static. We are not a bundle of unchanging qualities. Identities might shift depending on where we are, or who we are with. Thus, the things that matter in the production of any given knowledge may depend on the research we conduct and people we are working with. We need to be alert to the fact that situating knowledge is not a task that can be 'done and dusted' with a simple reflection on who we are, but is a method we must enact throughout our research in order to follow the complex politics of positionality that can unfold during the process.

The third point Rose highlights relates to the role of others in shaping our research. Human geography research often involves working with people (even when we work in an archive, we are arguably still working *with* people, we are just working with the traces they have left behind in memos, ledgers, diaries and so on). Consequently, when we construct knowledge about the world through research and writing, the people we encounter along the way are not external to the process. Like us, they are fully embroiled in the research and it makes sense that they shape it too. Reflexivity requires us to look outwards, not just to 'ourselves and our place in the world' (to follow Moss, 1995) but also to the positionality of our participants to attempt to gauge the way they inform what we know and how we know it. As Gillian Rose notes right at the start of her article (1997: 305), making sense of her research was as much about considering her own positionality as trying to make sense of that of her interviewee. It is tricky enough to practice reflexivity in light of our own role in making geographical knowledge, let alone the role of those we work with. But it is helpful to understand that research is an **interaction** and it is those communications between the researcher (themselves positioned and placed) and the researched (who likewise have a position and placement) which produce particular forms of understanding that are situated in that context. In this way, all knowledge is not only influenced by people and places, but also by time: the time the research took place, the events that were happening at that moment, and the factors that collided together in that instant to create the conditions in which the knowledge was produced.

The final major point that Rose urges us to consider is the intrinsic difficulty of actually doing reflexive research. Reflexivity, she notes, seems destined for failure (1997: 305). How can we ever really grasp the difference we make to the research we conduct? How can we assume that one 'facet of self' (Rose 1997: 306) over another will or does matter? How can we begin to assess the part our participants play in shaping how we react to them and them to us? How can we gauge whether or not the context of our research and the time it took place has had an impact on the kind of knowledge that results? Surely all of this is simply too tricky to contemplate, let alone to try and enfold into a workable method we can deploy to create more accountable research? Yet as Rose demonstrates, in spite of any failings that might come hand-in-hand with doing reflexive research, we shouldn't just dismiss it. As she notes:

> We cannot know everything, nor can we survey power as if we can fully understand, control or redistribute it. What we may be able to do is something rather more modest but, perhaps, rather more radical; to inscribe into our research practices some absences and fallibilities while recognizing that the significance of this does not rest entirely in our own hands. (Rose, 1997: 319)

Sam Saville's reflections demonstrate how we can engage meaningfully with reflective practice (see the Graduate Guidance box below). The following final section of this chapter deals with some of the practicalities of doing research reflexively, highlighting the considerations we can make for doing this altogether more 'modest' yet 'radical' form of research.

SITUATING KNOWLEDGE IN SVALBARD
SAM SAVILLE

My research explored the concept of value in Svalbard, in the high Arctic. Over the course of three field trips, I investigated the relationships between the human societies, material landscape and non-human, or 'more-than-human' life there. I used a combination of research methods to collect my data: interviews, participant observation, questionnaires and focus groups. Alongside this, a research diary, photographs and audio recordings helped me to make further observations and reflections.

I realised that I would be bringing my own values to this research very early on. The project would involve me flying to Svalbard and this was a serious consideration. Previously, I had been working for an environmental education charity and I had boycotted flying for seven years to reduce my carbon footprint. Field trips to the Arctic meant both engaging with and avoiding debates surrounding climate change linked to this region. Yet I was also confronted with further (unexpected) challenges to my own environmental ideals when I interviewed participants about their values.

The hunting and trading of animal skins is one example. In interviews where this came up, I realised my line of questioning had been affected by my own values on the topic. Like all of my research, interviews would have been done differently by someone else, who would have likely asked different questions, eliciting different answers.

(Continued)

(Continued)

In such situations I found that my research diary provided an outlet for suppressed personal opinions on the matter, and the material I recorded here became useful in recognising my position throughout the analysis and interpretive stages of my work.

Other surprises had a more drastic effect on the shape of my research. At the beginning of my project I assumed that climate change would be a matter of great concern to the residents of Svalbard. I therefore thought it would be central to my project. However, I had to accept that not everybody was so focused on climate change, or necessarily agreed it was an issue at all. Hence, whilst still very much focused on the relationship between humans and the environment and the value-making processes between the two, my project did not maintain climate change as the centre piece of the research.

I was also very aware of being the 'outsider' in Svalbard's small, close-knit towns. In a highly political environment where, as one participant noted, 'everything everyone does on Svalbard, including the janitors, is part of a geopolitical framework' (Interview 12, 27 May 2014), how I was perceived by the communities I was researching was important. In order to learn about the full range of value processes occurring in Svalbard, I needed to both suppress and enhance my own identities at different stages and with different groups. Whilst never concealing my role as a researcher, simple acts such as putting my camera away when wanting to blend in with the 'locals'; being an over-active photographer when I was with tourists or the photography club; or keeping environmentalist outbursts in check, helped me to gain a large set of data from a wide range of perspectives.

Placing yourself in your research

Being aware of the politics of knowledge production is the first step to doing the kind of research that acknowledges the partiality of the collecting and analysing data. However, translating an awareness into practice is altogether more difficult. The following suggestions map out ways in which we might do reflexive research successfully.

We must take reflexivity seriously

If reflexivity is a 'goal' of critical geographical scholarship (Rose, 1997: 306), then it makes sense that we should all be doing it. But recent arguments have stressed that we shouldn't just 'do' reflexivity for the sake of trying to prove our research is somehow more considered, ethical and accountable. If we don't 'do' reflexivity properly, it serves little to no benefit to the research we produce. Lynch provides such a warning in his critique of reflexivity (2000). He notes that reflexivity has claimed a status as some kind of 'virtuous' approach to make research processes and knowledges more transparent. Accordingly, scholars are simply giving lip-service to issues of positionality rather than taking seriously what this means for their research.

For example, it can seem simple to assume that your position as a woman, as middle class, as white, may shape your data. It can seem easy enough to write this into your dissertation and leave it at that. Accordingly you have 'ticked a box' and shown your awareness, and claimed to situate your knowledge. But how useful is such a statement? As Lynch argues:

> There is no particular advantage to 'being' reflexive, or 'doing' reflexive analysis unless something provocative, interesting or revealing comes from it. … Depending on the case, it may come across as insightful, witty, convincing (or) unconvincing, boring or silly. (2000: 42)

Thus, reflexivity is not just a 'throw-away' sentence to be included in your work to show you understand the politics that have shaped how you have done your research. If you are really serious about those politics, your approach needs to be altogether more considered, in offering a perceptive and critical assessment of how your research practice has been shaped by a host of positions and placings (see the Graduate Guidance box above).

We must critically consider our position

This means carefully and thoroughly considering issues of positionality beyond the obvious. For Jesse Heley, typically 'facets of self' such as gender, age or nationality were largely irrelevant. What mattered most was his position as local, and the shifting perceptions of this 'localness' to those he worked with (2011). When we consider what might shape the research we do, we need to be open to a whole host of attributes we hold, and to the wide range of settings we work in, to gauge how these might contextualise the knowledge we create. The following considerations in Information Box 6.1 can be helpful in critically engaging with issues of positionality during research.

INFORMATION BOX 6.1
POSITIONING YOUR RESEARCH

- Your positionality as a researcher may be defined by key identity characteristics (race, gender, sexuality, class). It may also be defined by other facets such as our previous experiences, ways of life, viewpoints and beliefs. Be open to the range of factors that 'position' you.
- Your positionality as a researcher is often defined in a hierarchy of power – you will find yourself either above or below your participants (see Sultana, 2007) and this can affect your research. Consider where you are situated in relations of power and how these

(Continued)

(Continued)

influence the questions you are able to ask, the access you are granted, the time you are given for research activities, and so on.

- Your positionality as a researcher is not static and fixed. Consider how you, yourself, change depending on who you are researching. Does your research require you to take on another persona (for example, might you put on a suit to interview someone 'elite', when you'd normally wear jeans and trainers? How does your appearance and demeanour with particular participants alter and how might this shape the research you collect?) Consider also how different parts of your identity become important at different times.
- Your positionality as a researcher is always made in relation to those you are researching. As Gillian Rose demonstrates, her positionality was constructed through her engagement with the Scottish, working-class man she was interviewing (1997: 306). Thus, our position is not a self-contained entity we take with us into the field. It is shaped by that place and the people we encounter there.

We must critically assess our methods

We shouldn't just consider ourselves and others when we research, but should also be critical of the methods we employ and the materials we engage with. How has our positionality shaped the questionnaire survey we have authored, or the interview script we have prepared (based on our own understandings, belief systems and values as to what is important to ask, and our assumptions of what we consider our participants will be able to answer)? How, in short, does reflexive practice help us to more critically consider the research tools we create or methods we use to gather data? (We consider these tools and methods in Chapters 7 and 8.)

Moreover, we might assume that such considerations are only vital if we are doing qualitative research – that is, research that seeks to generate subjective, partial and discursive knowledge about the world (see Chapter 2). Being reflexive seems at odds with quantitative methods that give us statistical, numerical answers from which we can make firm claims. Numbers are not disputable. They can give us factual answers. Surely they cannot be shaped as other forms of data might (interview, focus groups and observations)? Yet as Mei-Po Kwan has shown, it is vital to consider the politics that underlie quantitative data practices (2002). The data we gather will be shaped by the limits of our data collection processes and the decisions we make regarding sampling. Our data will be influenced by the test we select to gauge levels of significance. Even when we use secondary data (such as the Census) we must acknowledge that this is far from neutral. Census data is typically understood to represent the shape of the nation, but it always excludes parts of that nation (people such as asylum seekers, the homeless, and so on, who are not recorded through such statistics).

We should keep a research diary

A good method of employing reflexive practice is to keep a research diary (see Task 1.2). Diaries have an immediacy to them. We can jot down things that strike us as important as they unfold. We can also reflect back on situations and events, and record our thoughts and feelings retrospectively (see Graduate Guidance, pp. 13–14 above). Therefore, diaries – kept well – can act as useful tools for identifying when questions of positionality arise. As we write our diaries we might become alerted to the fact that a particular research exchange or encounter we had was shaped by a set of circumstances or positions. Or we might look back over our diaries and recognise a larger, more structural set of conditions (repeated issues of class or gender for example) that have impacted upon our data collection and interpretation. Task 6.2 dovetails with the previous diary task by providing a set of questions you may wish to consider in your diary/research blog during the process of doing your research.

TASK 6.2

Whilst Task 1.2 required you to keep a diary of your research journey as you design, do and deliver your human geography dissertation, Task 6.2 aims to provide a number of specific provocations to encourage you to think critically in your diary about your research as you do it. A successful diary is dependent upon 'deep' considerations of events, moments and occurrences; and your engagements with people, places and politics. Writing a diary means being prepared to question yourself and to not be embarrassed to commit your thoughts to paper. This isn't always easy, especially if you haven't written a diary before. The questions here are designed to ease you in to this process. Try to avoid short, one-word answers. Write in prose, and once you have begun, let your writing take you in any direction it leads.

- How did you find the process of interviewing, conducting surveys, engaging in observations? What did you think worked well? What would you change?
- Did anything particular happen that struck you as significant? Was anything particularly mundane or dull?
- What was the most interesting thing you found out from the interview, the focus group, etc.? Why was it interesting?
- What surprised you? (What didn't surprise you?) Why?
- Were there any moments where your role as a researcher became particularly apparent? If so, why? Do you think it mattered to the research you conduct?
- How were your engagements with participants? Were they relaxed, or uneasy, awkward or comfortable? How did this shape how the research progressed?

Keeping a diary is often considered to be part and parcel of good, reflective/reflexive research practice. Indeed, as Ian Cook notes (who has himself produced some excellent, reflexive accounts of research):

> ...researchers have been urged to write about their involvement in their own research because they may or should feel that they have to try and make sense of the tricky circumstances in which they studied before claiming to know anything about what they have studied. (Cook, 2001: 103)

Placing yourself in your research and reflecting on the significance of this placement is crucial (Rose, 1997: 306). A diary helps you to make sense of the world as you are constructing knowledge about it in your human geography dissertation. In the chapters to follow, attention to the ways we actually 'do' research is continued, with an overview of the various methods at your disposal for studying the spaces, places, people and events that are the focus of your study.

Chapter Summary

- Human geographers tend to write in the first person when relating their research findings. This reflects an acknowledgement that they are not separate or distanced from the research they conduct. Scholars now largely accept the role they play in shaping the data they collect and interpret, accepting that knowledge can therefore never claim to be neutral because it is always the product of the person who produced it.
- This idea dispels what is called the 'god trick'. Feminist scholars have demonstrated that no knowledge is free of the forces that have shaped it. Breaking the 'myth' of the god trick allows us to be receptive to a host of other voices and perspectives which create different knowledge(s).
- As researchers ourselves we must be attuned to our own role in making partial geographical knowledges. This means we must consider our positionality and how shifting 'facets of self' shape the work we do. This means critically considering our position to offer a more reflexive research account. In turn, this means situating our knowledge within its limitations and biases.
- Gillian Rose's key article helps us to understand the complexities of 'doing' reflexive research. However, she also notes that this effort can lead to failure. To ensure the best chance of success we need to embrace reflexive practice seriously (paying more than just lip service to our positionality). This means critically considering our position and our research methods. One effective way to do this is to keep a reflective/reflexive diary.

Key readings

Heley, J. (2011) 'On the potential of being a village boy: An argument for local rural ethnography', *Sociologia Ruralis* 51 (3), 219–37.
Lynch, M. (2000) 'Against reflexivity as an academic virtue and source of privileged knowledge', *Theory, Culture & Society* 17 (3): 26–54.
Moss, P. (ed.) (2002) *Feminist Geography in Practice: Research and Methods*. Oxford: Blackwell (see Chapters 6, 8, 9 and 11 in particular).
Sultana, F. (2007) 'Reflexivity, positionality and participatory ethics: Negotiating fieldwork dilemmas in international research', *ACME: An International E-Journal for Critical Geographies* 6 (3): 374–85.

7

MAKING RESEARCH HAPPEN: THE METHODS GLOSSARY

CHAPTER MAP

- Mapping out methods
- Research with texts and archives
- Research with people and places
- Research with numbers and maps

Mapping out methods

With the publication of the book *Military Geographies*, Rachel Woodward argued for the discipline to take the spatial experiences of military personnel, the formation and reformation of militarised landscapes, and the gendered geographies of military practice in training and in the theatre of war seriously (2004). Geography, Woodward argued, has been intimately and intrinsically shaped by military activities – including periods of short-term and sustained conflict, as well as a range of 'non-conflict' actions (for example, drills, manoeuvres and exercises). Moreover, military life is spatialised in countless ways. Military operations use space strategically for combat and for camouflage (Forsyth, 2013); military families migrate with personnel from place to place (Hyde, 2015); mobile technologies are used to target enemies from above (Adey et al., 2013), and military architecture (such as bases, bunkers and minefields) have transformed geographical terrains (Woodward, 2004). In short, Woodward's research has sought to answer the following question: what is the relationship between the military, space and place?

In order to answer any research question (such as the one above) geographers need to employ methods. Methods are integral to 'doing' research. They are the means by which data is collected to answer a given research query. As Alan Bryman describes:

> A research method is simply a technique for collecting data. It can involve a specific instrument, such as a self-completion questionnaire or a structured interview schedule, or participant observation whereby the researcher listens to and watches others. (2004: 27)

For Woodward's research, several techniques were employed to collect data. In order to understand the relationship between the military, space and place, interviews with personnel were conducted and soldiers' memoirs (in other words, secondary sources) were carefully analysed. Using these methods gave Woodward a deep insight into military spatial experience. However, before we can hope to use methods (as Woodward has done) we first need to know what different methods are available for geographical research.

In this chapter we consider the various techniques that are in the 'toolkit' for human geographers embarking on independent projects. It is important to read about methods as well as to practise them. Reading helps us gain a sense of the different types of methods that exist, their function, the sort of data they can elicit, their benefits, their limitations, how they can be used well, and the considerations you might make in employing them. This chapter offers an overview of the 'classic' methods often used by human geographers (Last, 2012). The role of this chapter is not to provide a definitive and exhaustive discussion of different techniques and their use (as other publications do this far more comprehensively; see Clifford et al., *Key Methods in Geography* (2010; 2016), for example). Instead the chapter acts as a summary of core methods. It provides a starting point that summarises a collection of commonly used methods alongside an individual reading list for each technique. You can also Go online! to access Sage journal articles that demonstrate the use of each method in practice. The chapter is split into three sections. The first deals with techniques for researching texts and archives. The second covers the methods typically used for researching people and places. The final section considers methods that involve collecting numerical or visually mapped data.

Go online! Visit **https://study.sagepub.com/yourhumangeography** for access to Sage journal articles that illustrate how each method highlighted in the chapter is used in practice. The first two papers demonstrate the use of textual and archival research; the second two show how interview, focus group and ethnographic data have been employed; and the final duo of papers outline the use of survey and mapped data in research practice.

Research with texts and archives

Stuart Aitken describes research with texts and archives as 'armchair' methods in the sense that you don't necessarily have to engage directly with people and places to accumulate data (2005: 233). Instead you might be able to access materials from books, catalogues, websites, newspapers, and so on. Often textual data is called 'secondary data' (see Information Box 7.1). It is often believed that to do 'good' research you have to go out and collect your own data (this is often called 'primary data'). However, there can be all kinds of pre-existing data that help in responding to specific research questions. In this section we consider textual analysis and archival research.

INFORMATION BOX 7.1
PRIMARY AND SECONDARY DATA

Although there has been some recent debate over the labels 'primary' and 'secondary' in relation to data sources (see Schutt, 2006), the two terms are commonly used to describe different kinds of information you can use as a geographer.

Primary data is that which does *not* pre-exist. It is data that isn't already out there, but which is created by the researcher (for example through interviews, focus groups, surveys, and so on). In short, primary data is the data you make – it is **self-constructed**.

Secondary data is that which has already been created by another individual, group or institution (for example government statistics, policy documents, archive material, maps, and so on). Secondary data is that which is **pre-constructed** – it is already out there for us to use.

Textual analysis

Textual analysis involves analysing a text (or number of texts) in order to grapple with how it shapes our understandings of space, place, time and movement. The texts that geographers can consider include conventional ones such as policy reports, newspapers and travel brochures. However, geographers can also examine visual texts such as photos, paintings, films, and even the physical environment (landscapes and architecture). It is useful to think about a text as anything that can be read that conveys socio-cultural and political meaning (Shurmer-Smith, 2002: 123). For geographers, texts are interesting research materials because they are not neutral, or innocent. Texts, in short, do things. They produce particular 'ways of seeing' (Barnes and Duncan, 1992). A wealth of research by geographers has demonstrated that texts are not just windows into the world through which we can better understand a given socio-spatial phenomenon – for example, global inequality, environmental harm, economic segregation – they are *constructs* of that world.

Texts are powerful in that they represent the world in a particular way, which can in turn provide a rationale for 'acquiring and subordinating [people and] space'

(Peet, 1998: 14). Let's take an example from the nineteenth century. During this age of imperial travel to the African continent, Felix Driver (2003) has noted how images were collected and then shown on lantern slides to audiences back at home in Britain. These images produced a particular narrative that 'stimulated patriotic effort and reinforced one's sense of place in the world' (Schwartz, 1996: 31). These were images used to justify the rule of distant lands because they constructed faraway people as exotic 'others' (Said, 1996). Now let's take an example from the twenty-first century. Consider the popular television show, 'The Only Way is Essex' (*TOWIE*). Entertaining as the show might be, it is powerful in constructing a particular spatially defined image about people from Essex. A former dissertation student I supervised conducted a detailed textual analysis to understand the way in which the programme constructed ideas about Essex life and society. The student had been brought up in Essex and was often upset by the associations that came with being labelled an 'Essex girl'. She was interested in the ways in which programmes such as *TOWIE* could repeat and perpetuate well-known stereotypes that 'placed' Essex residents as inferior, 'stupid' or 'common'.

When conducting textual analysis students should be rigorous and systematic in identifying and collating texts to analyse for their projects. Take the *TOWIE* project as an example. Whilst the texts were simple to access (via DVD and online collections) there were far too many episodes to analyse in the time available to complete the dissertation project. Accordingly, the student had to take a systematic cross-section of episodes, from different series, justifying this as a representative sample. Indeed, textual analysis is not an 'easy' method because texts tend to be readily available to us. Students should be able to convincingly state why particular texts have been consulted and not others. Accordingly, we must show we have searched out and selected texts relevant to our study. For example, take a project concerning the geographies of body size and shape (see Colls and Evans, 2014) and their representations in the media. Media sources – from webpages, magazines and so on – should not simply be 'cherry picked' because they look interesting. Rather we should take an approach where we look at a series of media representations from the same publications – or from a number of different publications – or from a set, specified date-range relevant to the study. In other words, data collection for textual analysis is not, and should not be 'ad hoc'. If texts are relevant to your study, contemplate how you may best collect and select them (and with online sources be sure to keep a record of the URL for later referencing). In Chapter 10 we discuss how you can analyse and 'make sense' of textual sources (see also Rose, 2012).

Key readings

Aitken, S. (2005) 'Textual Analysis: Reading culture and context', in R. Flowerdew and D. Martin (eds) *Methods in Human Geography: A Guide for Students Doing a Research Project* (second edition). London and New York: Routledge. pp. 233–49.

Doel, M. (2010) 'Analysing cultural texts', in N. Clifford, S. French and G. Valentine (eds) *Key Methods in Geography* (second edition). London: Sage. pp. 485–96.

Pink, S. (2012) *Advances in Visual Methodologies*. London: Sage.
Rose, G. (2012) *Visual Methodologies: An Introduction to Researching with Visual Materials* (third edition). London: Sage.

Archival work

If your research question requires you to understand and make sense of people, places and events you can no longer access, it is likely that you will need to look in an archive to piece the story together. Archives are depositories of information from the past (and by the past, this can be anything from 10 years ago to 500 years ago, or longer). Archives began to emerge systematically in the nineteenth century as governments sought to protect national histories (see Ogborn, 2003). Thus, archives came to be the 'official' depositories of times gone by, curating various events in particular ways (indeed, all archives are political, not everything is kept, and some documents, and therefore histories, are discarded – see Ogborn, 2003: 13). Today there is a wide range of archives that are at the disposal of geographers who are seeking to unlock geographical stories of the past.

First, there are National Archives that contain 'evidence of national culture and memory' (Hannam, 2002: 114). These archives contain government records, war records, records of public bodies, and so on. The UK has National Archives in England, Northern Ireland, Wales and Scotland, and most other countries will have their own equivalent. Archives of national importance can also be found in libraries. For example, the National Library of Scotland holds the records of its important nineteenth-century publisher John Murray including his notes, correspondence and ledgers (see Withers and Keighren, 2011). National libraries often hold materials that are relevant to state history, but are not government or public records.

Second, there are many archives on a local scale. If your research refers to a specific local area, it might be an idea to look in a local, city or county archive (most will hold old newspaper records, family records and civil proceedings on microfiche). Third, there are public and private archives which hold information related to a range of institutions or businesses. The BBC, for example, has an extensive archive. If your research relates to a particular organisation it is worth seeing if they have their own collection. For example, Sarah Mills' excellent work on citizenship and volunteering uses the extensive material held by the Scouting Association to make sense of how this particular body functioned as an educational institution in shaping youth for the future (Mills, 2013).

But how on earth do you find a relevant archive for your study? Luckily there is a host of databases for helping to identify where different material is held. Once you have your topic in mind you can search the databases and see what results are found. However, like all search engines, be aware that what you find will depend on the search terms you use (see also Chapter 4). Therefore, do not assume that there is no material if your search returns 'no hits'. Amend your search terms to see what might emerge. In a recent archival study of convict ships that facilitated migration from

Britain to Australian outposts, I and my colleague Jennifer Turner used various search terms – 'convict ships', 'prison ships', 'transportation' – each unearthing a range of relevant documents for us to look at (see Peters and Turner, 2015).

Go online! Visit **https://study.sagepub.com/yourhumangeography** to find a web link to The National Archives database which has contact details for over 2,000 archives.

Once you have found a potentially useful archive through a general search, go to the specific archive website and check if they have the material relevant to your research project (note, however, that some more specialist archives might not have a website so you may have to phone or email them to see if they hold the material you need). Most archives will allow you to search and request material online, even if the material itself is not available online (although some archives are in the midst of digitisation programmes). Thus, you often have to go to the archive itself to read documents (so make sure you can get to your archive – see Chapter 5). It is worth noting that archive work isn't often quick, or simple. You can never quite know what you might find until you arrive. It might be that the documents you find are not helpful. It could be that it takes lots of sifting to find what you are looking for. Indeed, archival material can take many shapes and forms, from handwritten notes, memos, letters and diaries, to maps, photos, drawings, statements, meeting minutes and so on. Befriending an archivist is often helpful. They are the experts and can help you to find the materials you are seeking.

Key readings

Baker, A. (1997) '"The dead don't answer questionnaires": Research and writing historical geography', *Journal of Geography in Higher Education* 21: 231–43.

Foster, J. and Sheppard, J. (2002) *British Archives: A Guide to Archival Resources in the UK.* Basingstoke: Macmillan.

Ogborn, M. (2003) 'Knowledge is power: Using archival research to interpret state formation', in A. Blunt, P. Gruffudd, J. May, M. Ogborn and D. Pinder (eds) *Cultural Geography in Practice*. London: Arnold. pp. 9–22.

Withers, C. (2003) 'Constructing "the geographical archive"', *Area* 34 (3): 303–11.

Research with people and places

Texts and archives often relate to people and places, yet we deal with them indirectly through the documents we encounter. If our research question isn't related to the distant past, or if we are able to access the people we wish to speak with, there are many techniques for engaging with people and places directly, accruing primary data. In this section we consider three key methods for researching people and places first hand: interviews, focus groups and participant observation.

Conducting interviews

If your research question seeks to find out the complex, intricate and subjective ways in which people and places relate, interviewing is an incredibly useful method. This is because it allows you to gain in-depth insights, compared to other techniques (say, questionnaires) where the respondent is only able to answer briefly, or within a fixed set of prescribed answers (Rubin and Rubin, 2005: 2-3). For example, in a project concerned with material culture and tourism, I have used interviews in my own work to hear people describe *in their own words* how objects from abroad have transformed their homes (see Peters, 2011b for a discussion of interviewing methods).

That said, whilst interviews are 'probably the most widely employed method in qualitative research' (Bryman, 2004: 319) their success depends on asking the 'right' questions, to the 'right' people, in the 'right' way. Typically, interviews tend to be categorised as **structured** (with a fixed set of questions to ask each participant), **semi-structured** (with a structure of questions and themes but with the capacity to go off on tangents or to ask additional questions), or **unstructured** (flowing wherever the conversation leads). Semi-structured interviews tend to be most common because they allow the conversation to remain focused around relevant research themes, but incorporate flexibility in allowing the conversation to flow naturally and change direction.

When conducting semi-structured interviews, it is usual to prepare a flexible interview schedule of questions for each participant, complete with prompts, should conversation stall (see Information Box 7.2, below). That said, as an interviewer, you must also be willing to deviate from your schedule, or to adjust your questioning depending on how the interview evolves. Given interviews are a conversation of sorts, they require you to listen carefully. It could be that you are discussing something emotional or difficult (research on flooding, for example, where people's homes or livelihoods were damaged) and you need to be thoughtful and sympathetic. It might be that you need to be clearer or more direct. Remember to always be respectful and to follow ethical guidelines (asking for consent and permission to use recording devices, see also Chapter 5).

Also think about the appropriate place to hold your interview(s). What location will be safe for you as a researcher? Will such a place enable you to hear your participant clearly, but also ensure they are not overheard (so confidentiality is protected)? Often, because of time and travel constraints, researchers will conduct interviews on the phone or via Skype. This adds a different dimension to interviewing as the personal touch in building rapport may be missing (as well as any technical issues in sound quality). Therefore, it is worth considering how to best operationalise your interview if you have to conduct it through the medium of technology (see Seltz, 2016). For example, ensure you are somewhere with good internet connection, or phone signal, and somewhere quiet where you will be able to hear properly.

Go online! Visit **https://study.sagepub.com/yourhumangeography** for access to the Sage journal article 'Pixilated partnerships, overcoming obstacles in qualitative interviews via Skype: A research note', by Sally Seltz in *Qualitative Research* (2016). This article outlines how to best conduct interviews using new online technology.

INFORMATION BOX 7.2
PREPARING AN INTERVIEW SCHEDULE

When preparing your interview questions, keep your central research question or problem in mind. What is your human geography dissertation seeking to find out? Ensure your interview questions allow you the best chance possible of gathering the necessary information. Most interview schedules will be divided into three sections, with questions for the beginning, middle and end of an interview.

The beginning

Start your schedule with so-called 'throw-away' questions to establish a comfortable dialogue. You might ask 'How are you today?', or pose an introductory question relevant to your topic: 'When did you first start working for the aid organisation?' These questions help to develop the flow of the conversation (and enable you and your participant to overcome any nerves). It is also an opportunity to build rapport with respondents. For example, during introductory questions you could provide some information about yourself, which often helps to build trust and respect between interviewer and interviewee.

The middle

Prepare a set of core or main questions which relate to what you want to know in order to answer your research question. However, be sure to listen. Sometimes the interviewee may say something that answers another question on your semi-structured script. Don't ask the question again as it will appear that you haven't been paying attention. Prepare prompts in case the conversation dries up. For example, you might ask interviewees: 'how did you achieve that?', 'can you explain why that happened?', 'what impact did that have?' and so on.

(Continued)

(Continued)

The end

Always ask the interviewee if they have anything further they would like to add, or if they have any questions you can answer for them as a researcher (although don't be afraid to say if you don't know the answer – you can always state you will find out and get back to them at a later time).

Always have a pen and paper with you, even if you are taking a Dictaphone to record the interview. This can be useful if your Dictaphone fails, but, importantly, it allows you to jot down anything of interest that might be mentioned mid-conversation that you want to follow up on later.

Key readings

Longhurst, R. (2010). 'Semi-structured interviews and focus groups', in N. Clifford, S. French and G. Valentine (eds) *Key Methods in Geography* (second edition). London: Sage. pp.103–15.

Seltz, S. (2016) 'Pixilated partnerships, overcoming obstacles in qualitative interviews via Skype: A research note', *Qualitative Research* 16 (2): 229–35.

Valentine, G. (2005) 'Tell me about... using interviews as a research methodology', in R. Flowerdew and D. Martin (eds) *Methods in Human Geography: A Guide for Students Doing A Research Project* (second edition). London and New York: Routledge.

Holding focus groups

Focus groups share some of the characteristics of interviews, as well as the skills necessary to conduct them (see the section on Interviews above). They are similar in that they elicit in-depth, detailed and conversational data. They also rely on the researcher asking carefully formed questions, often in a semi-structured style. Considering the guide to preparing questions in Information Box 7.2 above will be helpful if you are planning to use focus groups. However, focus groups are technically different to carry out. If you are seeking to find out varied opinions on a topic, or the views of a group in relation to your research problem, a focus group might be a useful approach. This is because, unlike the interview (which usually consists of just the researcher and the participant), focus groups bring together an ensemble (usually six or more) of participants who engage with the researcher and, crucially, with each other, around a particular theme or line of enquiry (Longhurst, 2010: 105).

Conducting a focus group requires careful consideration of recruitment, location choice, mediation and recording. Starting with the first, as with all research, you need to recruit participants who are relevant, choosing them 'on the basis of their experience related to the research topic' (Longhurst, 2010: 108). You can usually recruit participants by assembling a group yourself, or recruiting a pre-existing group (for example, a collective of anti-fracking campaigners). It could be, however, that no group exists (one, for example, that might share thoughts on sustainable cycle policies). If there is no pre-existing group, you will likely have to recruit and assemble a group of people yourself.

It is worth noting that the dynamics between pre-existing and assembled groups tend to differ. Pre-existing groups tend to already trust one another and, as such, focus group discussions can take less time to 'warm-up'. Participants of a group assembled by the researcher typically lack familiarity with each other, and may not want to express their opinions, beliefs, thoughts and experiences in front of others. This requires careful mediation (see below). Additionally, it can often be difficult to recruit enough people for a focus group and to arrange for them to meet in the same place at the same time. Thus, you have to be highly organised and flexible.

In terms of location you must think about your own safety as a researcher and meet in a suitable, public place. However, you should also consider meeting in a place that will be beneficial for the research you are conducting. Pick somewhere that you and the participants can reach easily (that is on a bus route, has parking, or is easy to find). On the one hand you may wish to pick somewhere 'neutral' – that isn't significant for you or the participants; on the other hand you may think it is useful to select a place the group usually meets (if they are a pre-existing group) in an effort to reproduce the social relations you are interested in understanding. You must also consider the way in which you arrange the room so that participants can see and talk easily with one another, but also so they don't feel exposed. As geographers we should be particularly attuned to how space matters to research practice (see also Chapter 9).

Once underway, whilst it is your job to pose questions to a focus group, the central job of the researcher is to act as a mediator to discussion. If the discussion is going as planned, you can just sit back and let it evolve – this is the ideal situation. That said, if the topic goes off course, it is your job to steer it back on course with appropriately timed entries to re-orientate the conversation. The mediator should also manage individuals in the group. Is one person talking over everyone else? Is one participant silent at the back? You often have to judge interventions carefully and with the right tone so as not to stifle the flow.

Finally, on a practical note, you have to think quite carefully how you record a focus group. Unlike an interview where a Dictaphone will often suffice, there is a drawback to this in a focus group research setting. If you use an audio recording it can be difficult, when listening back, to identify who is talking and to differentiate participants from one another. Often people will talk over each other too. It can be beneficial to video record focus groups, but you must ask permission and appreciate that the presence of a camera can often alter the dynamic.

Key readings

Bosco, F. and Herman, T. (2010) 'Focus groups as collaborative research performances', in D. DeLyser, S. Herbert, S. Aitken, M. Crang, and L. McDowell (eds) *The Sage Handbook of Qualitative Geography*. London: Sage. pp. 193–207.

Breen, R. (2006) 'A practical guide to focus-group research', *Journal of Geography in Higher Education* 30 (3): 463–75.

Stewart, D. and Shamdasani, P. (2015) *Focus Groups: Theory and Practice*. London: Sage.

Participant observation

If we want to understand a culture, community or a context, we might think to observe it, watching what people do and how places operate. This is typically called non-participant observation. However, we can also observe people and places by going one stage further and actually involving ourselves in the practices people are involved with and the places they are situated in. This is called 'participant observation'. At undergraduate level, we are most likely to conduct participant observation with communities or in places with which we already hold a connection, especially given the time constraints of completing the study. This requires us to 'step back' and make critical observations through our participation in activities and events with a given group. Participant observation, however, also shares traits with 'ethnography' (a term you will also come across in respect of this method). Ethnography is a form of participant observation conducted by outsiders to a given community and studies often take place over a long period of time (Brewer, 2000). Given the interests of human geographers in the relationships between people, space and place, such methods have offered much potential to researchers. Examples include Paul Simpson (who has participated as a busker in order to understand their spatial practices and experiences, 2011); Stephanie Merchant (who took part in underwater diving expeditions to understand the geographies of the body and embodiment in this setting, 2011); or Philip Crang (who conducted ethnographies of consumer culture in a fast food restaurant; see 1994 for a classic example). Participant observation is an excellent method for gaining 'close' and 'intimate' understandings of people's relationships with place but key considerations should be borne in mind when trying to conduct this method.

The first relates to access. If you are seeking to participate in the activities associated with a group of people, you will need to negotiate permission to enter the community in question. Whether you are familiar with the community you wish to research or not, you will need to consider the points raised in Chapter 5 regarding access. Participant observation can often require trust to be built up between the researcher (you) and the participant. Trust can take time to establish and time is crucial to participant observation. In order to fully understand how a community operates or how activities function (even in a setting we already have a knowledge of) we can't make a single observation and then claim to know everything. That said, 'how long it takes to become competent in what you choose to study, and indeed whether it is possible to reach that state, will vary according to what you choose to study using this method' (Laurier, 2010: 119).

Furthermore, participant observation requires a consideration of positionality. This is especially true because the researcher will straddle two positions when conducting this kind of research. They will be both **insider** and **outsider.** In participating with a group of people they (you) will cross a boundary and become part of that group. All the while you will still be an outsider because you are a researcher and only present in that role, in that setting, because of your project. Ethnography necessitates a reflexive approach (see Chapter 6), considering how you shuttle between these roles and how this shapes the knowledge you produce about the people and places you work with.

Ethics are also vitally important when conducting participant observation. Researchers often have to consider whether they conduct their observations overtly (where other people are aware you are completing the research) or covertly (where they do not). As researchers, it can often be tempting to consider carrying out observational work covertly. This is because we might assume that the behaviours of a group of people will change with the researcher present, compromising the quality of the data we collect and its ability to illustrate the ways of life we are seeking to understand. That said, for the most part it is best practice – in light of ethical considerations – to be overt when conducting participant observation.

When you conduct participant observation, the notepad becomes a very important research tool. Ethnographers record their observations by taking notes that are rich and descriptive (read Bradley Garrett's in-depth study with a group of Urban Explorers (2011a) for an example of what detailed ethnographic writing should look like). Typically researchers will make jotted notes whilst they are 'in the field' and they will consolidate these later. However, it is important to be aware of when you take notes (see Graduate Guidance, pp. 99–100). Ask yourself if it is appropriate to use your notebook if you are in the middle of an activity or event. It is also important to acknowledge that field notes won't work for all types of research (just think about Stephanie Merchant's underwater expeditions – the notepad would get wet). Accordingly, cameras and video cameras are now increasingly used for ethnographic research (see Garrett (2011b) and Spinney (2011) for a discussion of how these can be employed).

Key readings

Brewer, J. (2000) *Ethnography*. Buckingham: Open University Press.

Garrett, B.L. (2011a) 'Cracking the Paris carrières: Corporal terror and illicit encounter under the city of light', *ACME: An International E-Journal for Critical Geographies* 10 (2): 269–77.

Laurier, E. (2010) 'Participant observation' in N. Clifford, S. French and G. Valentine (eds) *Key Methods in Geography* (second edition). London: Sage: pp. 116–30.

Research with numbers and maps

Interviews, focus groups and participant observation are a few ways to grasp socio-spatial relations. However, numerical and mapped data are also important for understanding the connections between people and place in certain kinds of human geography projects. Sometimes numerical and mapped data are used as a 'context' to dissertation projects, providing necessary background information (White, 2010: 68). However, quantitative data collection can also be central to human geography projects (Dorling, 2010). This section covers some of the classic methods associated with numbers and maps that are central to human geography research.

Using survey data and other 'big' data sources

In 2011, the Office for National Statistics (ONS), the agency responsible for national data capture in the UK, began the first survey on personal well-being (dubbed the 'happiness survey'). In 2015, a combined dataset of three years of results was made available to the public for the first time. The survey results contain data – on a national scale – on matters of life satisfaction, happiness, anxiety and feeling of worth. The data can be broken down into different regions and demographics (age, sex, employment type, and so on).

Pre-existing, or secondary data sources (see Information Box 7.1, p. 108 above) of already-compiled statistics can be an excellent resource for human geography projects (see Dorling, 2010; Kitchin and Tate, 2000; White, 2010). If your research question seeks to uncover the statistical relationships between different experiences, attributes or expectations (happiness, income, mortality rates) and specific places (local, regional, national) then this kind of data can provide a fantastic context or can provide key data to analyse in respect of your research enquiry.

Go online! Visit **https://study.sagepub.com/yourhumangeography** to find a web link to the UK Data Service which holds a range of sources that could be used in student projects, including UK surveys, cross-national surveys, longitudinal studies, Census datasets (to name but a few).

But why would you use this kind of pre-existing data and not collect it yourself? Let's take an example. If your dissertation sought to find out if people living in different counties of Wales experienced different levels of educational attainment, you could conduct a questionnaire survey yourself – sending it out to lots of people in different regions across Wales. However, this would likely be time consuming, expensive and return rates on the survey would typically be low (see McLafferty, 2010). Moreover, this data probably already exists via national surveys conducted by the government. If your research problem considers a 'big' question in respect of scale you can often use what is called 'official data' in your studies to answer research questions. Paul White (2010: 61-76) states the benefits of this kind of data: it often covers wide geographical areas; it tends to be free for us to use; and it is typically collected using robust and thorough techniques.

However, it isn't just 'official' government statistics that can provide large-scale datasets that we can employ in our projects. We can also use non-governmental data (MORI opinion polls, or Gallup election polls, for example). Whilst these surveys (official and unofficial) allow us access to raw statistical data to use in our projects, it is also worth noting that there is an increasing range of 'big data' that geographers can now use to understand the social world (which isn't just restricted to survey data). Big data is that which is high in volume (in other words there is a vast array of data, often electronic), high velocity (captured in real-time) and exhaustive (wide in scope, often covering large geographical areas) (see Kitchin, 2013: 262). Big data often appears in raw form (in the sense it is 'just' numbers, that haven't yet been analysed). It can include information on the number of users of a website, or phone app (for example, the numbers of 'hits' a page receives, or the volume of ticket bookings for global flights). It can also include biometric data that is captured from technological devices used for running or cycling that may be uploaded to specialist websites and databases (see Kitchin, 2013: 263). Moreover, 'big data' is not only numerical. We might also consider 'big' qualitative datasets that exist through social media outlets (such as Twitter) or via online reviews, which give us an extensive dataset of subject opinions (such as TripAdvisor).

It is worth noting that these kind of datasets (unofficial and 'big' datasets) are not always free to use, may be more difficult to find, and the methods by which data has been collected may not always be as transparent. As Rob Kitchin notes (2013: 263), we are now in the midst of a 'data deluge', and whilst this provides geographers with opportunities, there are also risks and challenges to using the data sources available. Like all secondary data we should consider how it has been shaped by those who have produced it and consider how we might go about making sense of it to arrive at an answer to our research problem (see Chapter 10).

Key readings

Fotheringham, S., Brunsdon, C. and Charlton, M. (2007) *Quantitative Geography: Perspectives on Spatial Data Analysis*. London: Sage.

Kitchin, R. (2013) 'Big data and human geography: Opportunities, challenges and risks', *Dialogues in Human Geography* 3 (3): 262–67.

White, P. (2010) 'Making use of secondary data', in N. Clifford, S. French and G. Valentine (eds) *Key Methods in Geography* (second edition). London: Sage. pp. 61–76.

Administering questionnaires

Whilst survey and 'big' data gives us access to a variety of information, pre-collected and often on a large scale, geographers also administer their own surveys or questionnaires when completing research. If you require statistical data, but at a more local level, regarding a particular topic or theme, you might choose to conduct your own questionnaire. As McLafferty notes (2010: 76–7), 'survey research is particularly useful for eliciting people's attitudes and opinions about social, political, and environmental issues such as neighbourhood quality of life, or environmental problems and risks'. However, if we choose to use a survey, we have to design it carefully so that it is able to generate the data we need.

Similar to interviews and focus groups (see the sections above) the questions that feature as part of a survey need to be considered and clear. However, questionnaires rely on a fixed set of questions. The same questions are asked, in the same way, in the same order, to all participants. Questions typically require short answers or for the respondent to select an answer from a specific list of responses. This allows the data to be compared. Typically, questionnaires will include the collection of demographic data (which can be used to conduct analysis which correlates different factors, for example age and attitudes to recycling) and the collection of data specific to the research question (whatever the topic might be). We can use a range of different questions to collect the data relevant to our study (Denscombe, 2007). Figure 7.1 sets out the different types of questions that can be used by researchers when constructing a questionnaire. Moreover, McLafferty provides a useful overview for wording your questions (Information Box 7.3).

INFORMATION BOX 7.3
WORDING FOR SURVEYS: DOS AND DON'TS

Do:

- keep questions simple;
- define terms clearly;
- provide instructions for how to answer questions.

Don't use:

- jargon or phrases people won't understand;
- long, complex, double-barrelled questions;
- biased or emotionally charged terms;
- negative words (e.g. 'not' or 'none' can lead people to tick/circle the wrong box/number).

(Based on McLafferty, 2010: 79)

The yes/no question

Do you use public transport (tick one box)

Yes ☐ No ☐

The agree/disagree question

Liverpool has excellent public transport options (tick one box)

Agree ☐ Disagree ☐

The list option question

Which of the following modes of public transport do you use? (Circle all that apply)

Bus

Metro

Train

Ferry

Taxi

The rank order question

Which forms of public transport do you believe offer the best value for money? Rank the highest three (1 being the best value, 2 being the second best value, 3 being the third best value)

Bus

Metro

Train

Ferry

Taxi

The likert scale question

Public transport in Liverpool is good value for money (circle one answer)

Strongly disagree Disagree Neither agree nor disagree Agree Strongly Agree

The rate item question

Please rate public transport in Liverpool in respect of the following factors (1 is poor, 5 is excellent)

	Poor				Excellent
Cleanliness	1	2	3	4	5
Safety	1	2	3	4	5
Value for money	1	2	3	4	5
Punctuality	1	2	3	4	5
Useful routes	1	2	3	4	5

The statement question

Please add any further comments regarding your experience of public transport in Liverpool:

Figure 7.1 Designing a questionnaire: survey question types. (Source: Kimberley Peters)

It is also vital, as McLafferty notes, for students to consider the type of analysis they hope to conduct when constructing the survey. Conducting a survey isn't about using every different type of question, but rather the ones suitable for gathering the information you require. For example, will your question types allow you to gather data that can be easily correlated with statistical tests? Kitchin and Tate (2000: 49-52) provide an excellent guide to question formulation for geographers seeking to generate their own quantitative data (see also Rogerson, 2014).

Once you have your survey, you should first pilot it to ensure your questions are easily understood and clear. Next you must decide how to administer it. This means considering your sample and the means by which you will circulate it. You can use a random sample (where your target population is selected at random), a systematic sample (choosing respondents at regular intervals) or a stratified sample (that represents various subgroups by dividing population and then sampling within stratified frames of gender, age, race) (see McLafferty 2010: 84-6). You must also consider your sample size. The greater your number of respondents, the greater confidence you can have in your data. A smaller sample will mean your data has more bias.

It is also essential to think about the practicalities of conducting the survey. Questionnaires can be carried out face-to-face, on the telephone, by post, by dropping them off door-to-door and picking them up, or administering them online. Each of these techniques has differing pros and cons (see McLafferty (2010) for an in-depth explanation). Online surveys are now increasingly popular (see Madge and O'Connor, 2002) because they are simple to circulate, cheap to conduct, have a good geographic reach, and often collate data automatically. However, they may not be suitable for all research projects (see Chapter 8).

Key readings

Kitchin, R. and Tate, T. (2001) *Conducting Research into Human Geography.* Harlow: Prentice Hall (particularly Chapter 3).

McLafferty, S.L. (2010) 'Conducting questionnaire surveys', in N. Clifford, S. French and G. Valentine (eds) *Key Methods in Geography* (second edition). London: Sage. pp. 77–88.

van Riper, C.J., Kyle, G.T., Sutton, S.G., Barnes, M. and Sherrouse, B.C. (2012) 'Mapping outdoor recreationists' perceived social values for ecosystem services at Hinchinbrook Island National Park, Australia', *Applied Geography* 35 (1): 164–73 (This provides a good example of how a survey can be used to collect data for a human geography project).

Engaging with Geographical Information Systems

Geographers have a long relationship with visual and mapped data (see Rose, 2003). Geographical Information Systems, or GIS for short, are specific technologies that allow us to 'visualize, question, analyse, and interpret data to understand relationships, patterns, and trends' (ESRI, n.d.). GIS are tools that allow us to store and analyse geospatial data (that is, data that is defined by being locatable on the Earth's surface) and visualise it in map form. There are a various software programmes that facilitate such analysis and your institution will probably subscribe to one for your own use in geography projects. Depending on the software available, researchers will be able to apply a wide range of techniques for the processing, analysis and management of geospatial data. In doing so, this allows users to visualise their data as well as perform quantitative analysis to interrogate datasets. But how do GIS technologies work?

GIS rely on mapped data, that is, data with a specific spatial reference. Typically, geographical data is digitally stored in one of two formats: raster (which is data stored in regular grids or pixels) or vector (which is data stored in irregular points, lines and polygons). Batty notes that most technology will now enable users to 'integrate' and 'move ... between' these different forms of mapped data for ease of use (2010: 410). However, whilst the type of data you use will shape the types of analysis you can apply within GIS, the more important function relates to attributes. Attributes are the characteristics attached to geographical features: population size, average income, elevation (Goodchild, 2010: 379); vegetation type, agricultural use, primary industry and so on. GIS become very helpful through their capacity to take attribute data (of various kinds) and produce layered maps that integrate data in mapped form. Furthermore, a wide range of functions can be used to analyse these attributes in a quantitative manner: for instance, to determine relationships between attributes, spatial patterns and hotspots that occur within datasets. For example, geographers have been able to use GIS to map the relationships between people, place and disease (Hardy et al., 2015). Here it might be possible to layer and integrate mapped data regarding vegetation and climate with average ages of a population in a place, with disease 'hot spots' (indicated by data charting the number of cases of a particular outbreak, for example). This then allows the user to see if there are patterns between these attributes.

GIS are not only helpful for carrying out spatial analysis but can also be used as a tool for predicting change over time (Batty, 2010: 412). Batty notes how population maps of London that are created using existing data plotted into map form can then be used as the basis for applying map algebra to generate 'future' predictions (see Batty, 2010). GIS can also be used to analyse spatial networks. Consider a standard SatNav in a car that plots ideal routes over networks of roads. Essentially these roads represent networks of vector line data with each line segment having associated attributes taking into consideration factors such as average speed, distance, presence of junctions/roundabouts in calculating an optimal route. Additionally, GIS may be used to understand the context of geographical phenomena by visualising knowledge of surrounding areas that may be a factor in the geospatial patterns present on any map produced (Goodchild, 2010).

Accordingly, GIS methods are now increasingly used by human geographers to understand – in mapped, visual form – particular geographical relationships between people and place. These may be quantitative relationships (such as population size, or mortality rates) but can also be qualitative relationships (see Cope and Elwood, 2009). As GIS technology develops, the capacity to produce maps that depict, and more importantly explain, geographical phenomena, are becoming more commonplace. As Aswani and Lauer have noted, in regard to their study of the role of local knowledge in establishing fishing zones around the Solomon Islands:

> GIS visual representation made spatio-temporal differences in human fishing patterns more apparent than would have been the case if only interviews or participation in people's fishing activities had been relied upon. (2006: 96)

In other words, GIS creates knowledge by 'making it possible to present data in a variety of useful ways' which then 'expose(s) the meaning in data that would not otherwise be apparent to an observer' (Goodchild, 2010: 379), or may not be possible through the use of other methods. In short, GIS does not just display data, it generates new data that can help us to understand the social world.

Key readings

Aswani, S. and Lauer, M. (2006) 'Incorporating fishermen's local knowledge and behavior into geographical information systems (GIS) for designing marine protected areas in Oceania', *Human Organization* 65 (1): 81–102.

Batty, M (2010). 'Using Geographical Information Systems' in N. Clifford, S. French and G. Valentine (eds) *Key Methods in Geography* (second edition). London: pp. 408–22.

Cope, M. and Elwood, S. (eds) (2009) *Qualitative GIS: A Mixed Methods Approach*. London: Sage.

Chapter Summary

- Human geographers can conduct research with texts and archives. A text is anything that can be 'read': a policy document, a film, a painting, performance, building or landscape. It is analysed to extract socio-cultural and political meaning. Archive research involves a consideration of records from the past in order to make sense of historical or contemporary geographical phenomena. Geographers can use a range of archive material held in national or local depositaries, or collated by public or private bodies.
- There is also a set of methods typically used for researching people and places that generate qualitative data. These include in-depth interviews, focus groups and participant observation. These methods require researchers to carefully consider the questions they ask; how they may negotiate access to participants or groups; the ways in which they record data effectively, and the means of conducting research ethically.

- Methods that involve collecting numerical or mapped data are also prominent in the repertoire of techniques used by human geographers. Secondary survey datasets are commonly used by geographers to uncover spatial relationships. 'Big data' is also becoming popular in geographical analysis. Geographers can also collect their own comparable data through employing questionnaires. Increasingly, use of GIS is a method that can be used to generate data that visually demonstrate the patterns of connections that exist between people, activities, events and place.

Key readings

Clifford, N., French, S. and Valentine, G. (eds) (2010) *Key Methods in Geography* (second edition). London: Sage.

DeLyser, D., Herbert, S., Aikin, S., Crang, M. and McDowell, L. (eds) (2010). *The Sage Handbook of Qualitative Geography*. London: Sage.

Fotheringham, S., Brunsdon, C. and Charlton, M. (2007) *Quantitative Geography: Perspectives on Spatial Data Analysis*. London: Sage.

8

MORE ON METHODS: APPROACHING COMPLEX SOCIAL WORLDS

CHAPTER MAP

- Introducing something different
- Participatory and action-orientated geographies
- Experimental and creative geographies
- Sensory and mobile methods
- Researching online

Introducing something different

It is early autumn 2011. Inspired by the latest uprisings in the Middle East, and driven by the recent rounds of austerity cuts and a growing inequality gap between the richest and poorest, a collective forms at Wall Street – the financial capital of New York. Soon this gathering settles at Zuccotti Park. The 'Occupy' movement is born. This was no ordinary protest – assembling, marching, chanting, disbanding – a fleeting, disruptive, provocative subversion to everyday life in the city. Occupy stayed true to its title. From its very beginnings it was a highly spatial phenomenon, involving the inhabitation of space. Protesters were to fill the street, and there they would stay, settled, occupying camps, a visible and constant reminder of their cause. And this was no isolated, rooted form of contestation against the capitalist structures that oppressed the '99%' of the population (as the slogan came to say) disadvantaged by systems of power that prioritised the wealthy. Whilst Occupy was certainly concerned with rootedness – with the occupation of space – it was also a 'movement'. Accordingly, this was not a static protest. Rather it unfolded via social networking and media technologies, creating a global web of occupations in key financial districts across the world. A month after protesters occupied Wall Street, camps began to establish themselves in London, outside St Paul's Cathedral in the heart of the financial city. Only a series of court injunctions could shift the protesters in the capital. Occupy was (and is) many things. People participated. They engaged in activism to provoke change. They settled, and spread, and moved. They creatively subverted plays of power to make a point (see Breau, 2014; Pickerill and Krinsky, 2012).

How, as geographers, can we use methods to understand this highly spatial event that occurred in major cities around the world? Angela Last (2012) notes that there is a set of 'classic' methods used in the social sciences and arts and humanities. We considered these in Chapter 7. Certainly we could use these methods to generate understandings of the Occupy movement (we could interview protesters, search textual records such as newspapers or map activist movements). But do these methods allow us to really get to grips with what was (and still is) happening – with the choreography of the crowd, the embodied emotion and sensation of protest, the purposeful action initiated? Events such as Occupy alert us that we might need to adapt how we do research when we are generating knowledge about a world that is made through action, performance, movement and social networks. This chapter aims to do something different, introducing a range of geographical approaches and techniques that develop, enhance and invigorate how we 'do' research, and that present us with novel ways of collecting data about complex social worlds (Shaw et al., 2015: 211).

In what follows, this chapter outlines four innovations in research practice that might be helpful as you 'do' your dissertation project. The first concerns participatory and action-orientated geographies that seek to renegotiate the relationships between researchers and participants, and which aim to challenge prevalent geographical worlds through activism. The second considers what are now termed 'experimental' or creative geographies, which are increasingly adding methodological novelty to how we conduct research (including drawing, knitting, poetry, dance and live theatre). Thirdly, the chapter highlights embodied, sensuous and mobile approaches that are assisting in understanding a world that is motionful and lived. Finally, attention turns to the internet and the capacity of online tools and technologies to provide new ways of researching and understanding the social world.

Go online! Visit **https://study.sagepub.com/yourhumangeography** for access to Sage journal articles that illustrate how each method highlighted in the chapter is used in practice. Here you can read about how human geographers have used participatory and action-orientated approaches; creative and experimental practices; sensory and mobile methods, and online techniques.

Participatory and action-orientated geographies

Chapter 1 of this book offered some ruminations on why we become geographers. Certainly, there is inquisitiveness. But there is also, often, a desire to engage with the world in order to not only understand it, but to change it, or to do some good that positively impacts particular places and the people who use them. As Gill Valentine has written, '[g]eographers have long been concerned with questions of social justice' (2003: 375; see also Ward, 2007). Accordingly, participatory and action-orientated geographies seek to 'do' things – to 'do good' (Wynne-Jones et al., 2015: 1) and to 'work towards change' (Pain and Francis, 2003: 46). Both are driven by a desire to make geography relevant and useful beyond the 'ivory towers' of the university. In addition, both of these approaches seek to challenge the 'top–down' power relations that shape much academic work (Breitbart, 2010, Pain and Francis, 2003). Typically, academic work is conceived by the researcher (me and you). The aim of participatory and action-orientated research is to hand this process back to those we work with, and to dissolve the 'researcher-researched distinction' with a ground–up approach (Ward, 2007: 6). Indeed, participatory approaches seek to 'fully engage' participants with the research process, 'democratising' and 'demystifying' academic work in the process (Breitbart, 2010: 142). Action-orientated approaches see the researcher integrate with the cause they are studying, taking a role at grassroots level (Pain, 2004). This breaks down the boundary between the researcher and researched, as geographers are activists themselves (see Ward, 2007: 4).

There is a range of examples of participatory and action-orientated geographies that demonstrate the application of such approaches in practice (see the excellent edited collection *Participatory Action Research Approaches and Methods,* by Kindon et al., 2007). Pain and Francis (2003: 47) note how much participatory work has developed in the Global South in order to challenge the power relations such research often reinforces. However, their research demonstrates the application of participatory approaches to the Global North as a means of constructing knowledge of how vulnerable young people experience crime and its role in shaping their life trajectories. Pain and Francis (2003) used a range of participatory methods – asking young people to create timelines of their lives and relationships with crime, and to brainstorm ideas on what they felt 'crime' meant – putting participants at the very heart of data generation. But participatory approaches are also being used more widely. Hilary Geoghegan's recent work engages with 'citizen science' where academics involve participants in research to understand how everyday knowledge about the environment is formed (see Geoghegan, n.d). Action-orientated work is less focused on integrating participants into the research process, than on integrating academics into research by positioning them as activists themselves. Examples of this kind of work include instances where researchers have engaged in Environmental Direct Action (EDA) – in other words involving themselves in protests, blockades, marches and sit-ins as a response to environmental and social issues (see Anderson, 2004; Chatterton, 2002) – and forms of online activism in response to ecological and political concerns (see Pickerill, 2003).

But how might you consider adopting a participatory or action-orientated approach in your own dissertation work? Such approaches tend to be encouraged from the very start of a project (the design stage, see Section I of this book). However, that said, research is a process and it is never too late to engage with participatory approaches. Indeed, there is no 'set' way to engage in participatory, action-orientated work (especially when we consider it should be driven by research participants not the researcher). Bradley Garrett and Katherine Brickell (2015) have noted there is often a 'gold standard' for participatory and action-orientated research – a way of doing such research to 'best effect'. Yet in their study of participatory videoing with communities in Cambodia around themes of domestic violence, they demonstrate the challenges of doing work that places communities at the centre of research enquiry. They problematise the assumption that research can better achieve its goals simply because it involves participants. Thus, we can (and should) be flexible in our approaches to participatory, action-orientated research, while staying true to the 'ideals' of this kind of research in standing back as researchers and letting our participants take the lead.

To do so, participatory and action-orientated approaches use methods that enable participant involvement and researcher integration. 'Classic' methods such as interviewing and ethnography remain important (the latter is particularly useful in activist research – see Jon Anderson's consideration of his own actions in respect of an anti-quarry campaign in Bristol, 2004) – but researchers can also use a host of other techniques, often those that transfer ownership of data collection to participants. For example, Garrett and Brickell demonstrate how understandings of women's lives in rural Cambodia could be better understood by providing cameras to participants, allowing them to construct their own narratives around the research theme (2015). Pain and Francis demonstrate the use of 'paper, coloured pens, sticky dots, post-its and other visual materials' as helpful tools for allowing young people to express their experiences (2003: 48). They also note how engaging participants in 'mapping, timelines, cartoons, matrices and pie charts' can assist in generating participant-led knowledge.

All this said, given a strong focus on global examples and ethically sensitive research subjects in much participatory and action-orientated research, as a dissertation student you might be forgiven for thinking that such approaches are unfeasible given the scale and time-frame of your project. However, student researchers should not be put off from engaging with this kind of research. Taylor's study of the Brixton Pound (2014; see also North (2006) on alternative economies), for example, demonstrates how such methods can be productively used in a range of novel projects. In this work, Taylor examines how activist groups have developed local currencies to foster community ties and bolster small economies. The 'Brixton Pound group' participated in Taylor's research through a range of research activities (such as interviews) and Taylor herself became 'active', involving herself in group meetings and events. The dialogue created between the researcher and researched created 'useful' interactions. For example, material collected by Taylor was used by the group to support their actions (for example, in applying for funds to develop the initiative). As such, Taylor's work identifies how participatory, action-orientated approaches may well be more subtle, but can nonetheless help stimulate change, in ways that are no less radical.

Key readings

Pain, R. and Francis, P. (2003) 'Reflections on participatory research', *Area* 35 (1): 46–54.

Ward, K. (2007) 'Geography and public policy: Activist, participatory, and policy geographies', *Progress in Human Geography* 31 (5): 695–705.

Experimental and creative geographies

In 1995, the late geographer Allan Pred published the ground-breaking book *Recognising European Modernities.* On the one hand, the text offered a theoretically perceptive foray into the making of capitalist modernity through a critical consideration of spaces of consumption and spectacle in Europe (specifically, Sweden). Yet the book also did something altogether more radical. To present his case, Pred constructed the text as a montage; as fragments of writing, carefully positioned and pieced together. Visually, the pages of the book seemed to break with normal academic writing conventions. Words were offset, the text embraced different font sizes (often mid-sentence), poetry and prose alternated, and images appeared as collage. Pred provided the following rationale for his approach:

> Through assembling (choice) bits
>
> and (otherwise neglected or discarded) scraps,
>
> through the cut-and-paste practices of montage,
>
> one may attempt to bring alive,
>
> to open the text to multiple ways of knowing,
>
> and multiple sets of meaning,
>
> to allow differently situated voices to be heard,
>
> to speak to (or past) each other
>
> as well as to the contexts from which they emerge
>
> and to which they contribute.

> Through deliberately deploying the devices of montage,
> one may attempt, simultaneously,
> to reveal what is most central to the place and time in question
> by confronting the ordinary with the extraordinary,
> the commonplace with the out-of-place,
> the (would-be) hegemonic with the counterhegemonic,
> the ruly with the unruly,
> the power wielders with the subjects of power,
> the margin definers with the marginalized,
> the boundary drawers with the out-of-bounds,
> the norm makers with the "abnormal,"
> the dominating with the dominated.
>
> (Pred, 1995: 25; original layout)

Pred's work might be considered both experimental and creative. It might also be considered as a forerunner to the 'experimental' and 'creative' geographies that have been shaping geographical work since the turn of the millennium. But, to borrow a line of enquiry from Harriet Hawkins (2015: 248): 'what do creative geographies do (and how do they do it) in response to both current methodological and conceptual questions within geography and wider concerns with how our research goes to work in the world?' Why, as human geographers, should we care about experimental and creative approaches that are both growing in popularity, and which are becoming 'prized' (Last, 2012: 718) as ways of grasping complex social worlds?

Experimental and creative geographies are both an intellectual project interested in testing the boundaries of accepted spatial knowledge, and a site for methodological innovation in order to facilitate that project. On the one hand, experimental and creative approaches matter because they help in making sense of the non- and more-than-representational worlds which we inhabit; worlds that are lived and experienced (see Lorimer, 2005). As Hawkins notes (2015: 248), creative approaches are 'a response to the discipline's ongoing orientation towards embodied and practice-based doings'. On the other hand, they have been celebrated for bringing methodological vibrancy to the discipline – presenting new ways of 'doing' geography (see Shaw et al., 2015). Indeed, experimental geography has a legacy situated in scientific approaches and the laboratory – where formalised experiments would often take place (Kullman, 2013; Last, 2012). However, contemporary experimental geography is less concerned with the 'traditional vocabulary' of experimenting (Last, 2012: 706) but instead comes to encompass a range of experimental and creative approaches that on the one hand 'push the limitations of current conventions' and, on the other hand, do so by generating a range of engagements (with art, theatre, music, poetry) that also employ a range of methods to this effect (drawing, acting, performing, writing) (Last, 2012: 708).

Indeed, geographers have 'got creative' in a number of ways (see Hawkins, 2015; Madge, 2014). More traditionally, they have engaged with creative 'works' (often representations such as paintings, sculpture, poetry and prose) using these as windows to generate geographical knowledge. Increasingly, however, geographers have become creative themselves (either through collaboration with artists and practitioners – see earlier section on participatory geographies) or through using creative techniques to capture and collect geographical data (Madge, 2014: 178). Accordingly, creative geographies 'adapt' current, classic methods such as interviewing or mapping, using them in new and dynamic ways (Kullman, 2013: 888), whilst also establishing a wider repertoire of research tools (including cameras, pens, pencils, watercolour paints, charcoal, cutting boards, threads, tracing paper, to follow Hawkins, 2015: 248) for gathering data. Crucially, these methods are not simply fashionable, avant-garde approaches for the sake of being inventive. In answering the question 'what do creative geographies do?' (Hawkins, 2015: 248), we can argue that they make possible new geographical understandings. They help us to comprehend the world in new ways, and more significantly can also intervene in making and shaping geographies (see Hawkins, 2015: 261; see also Price, 2015: Rogers, 2012).

Let's return to Allan Pred's writing project (1995). Montage was not simply a tool for aesthetic novelty. It served a distinct purpose in generating meanings and constructing knowledge of the industrial and high-modernities under examination. In other words, as Hawkins argues (2015: 264), experimental and creative practices have a hand in uncovering geographical knowledge that would otherwise be unintelligible. Creative approaches aren't just a means of 'better' (re)presenting geographical research to multiple audiences (other scholars and the public). They are methods that help provoke, cajole, re-orientate and reposition what is known about the world (Last, 2012: 708).

Consider Laura Price's work on knitting practices and the city (2015). We might think of knitting as a banal, everyday, albeit uninteresting activity. Yet as Price demonstrates, knitting is a creative practice which forges communities, bringing networks of people together in knitting clubs or in craft stores. In this sense knitting produces spatial connections. Furthermore, knitting has been used in subversive ways through 'yarn-bombing' (a form of woven graffiti, where knitting is positioned around the city to intervene in the landscape). Here knitting comes to craft specific geographies, temporarily reshaping the masculinised city with materials of femininity and comfort (Price, 2015). Or consider Amanda Rogers' work on the performing arts and contemporary south-east Asian theatre (2012; 2015). Rogers demonstrates how theatre spaces work as sites for performing geographical identities while also challenging assumptions of those identities. In her work she demonstrates how the 'staging' of theatre productions can work to promote diversity or to entrench differences (2012: 70). Accordingly, creative approaches enable social geographies to be recognised and understood in new ways. Finally, consider the work of Clare Madge (who we will also return to later in the chapter). Madge uses poetry as a creative method of grappling with geopolitical relations (2014). Madge's own poetic writings become a way of constructing geographical knowledge about global conflict. She contends that although a complex medium – which we must use critically as a method – it enables a powerful expression of geopolitics to be realised (Madge, 2014: 181-2). As she notes, 'poetry has potential to give insight into the multiple (and sometimes painful) realities of life – not some emotionally-flattened version' (Madge, 2014: 181).

Go online! Visit **https://study.sagepub.com/yourhumangeography** for a web link to the journal *Cultural Geographies*. Browse the wide range of creative and experimental studies shaping work in the discipline in the 'Cultural Geographies in Practice' section.

Crucial to the work of Price, Rogers, Madge (and others) is their active engagement with creative methods to generate such knowledge. These geographers knitted, worked with theatre companies, wrote poetry. This begs the question of how we might use creative and experimental methods in our dissertation work, and if we are skilled enough to do so (following Hawkins, 2015). Creative and experimental approaches won't be for everyone. However, some dissertation projects may find it useful to engage with more novel methods. Ask yourself if your project might be enriched through engaging with alternative forms of data collection, and do not worry about lacking the relevant expertise (in painting, writing, sculpting, and so on). In her agenda-setting, interdisciplinary work that merges art, aesthetics and geography, Harriet Hawkins reflects upon her 'ineptitude' as an artist, and her inability to accurately record her surroundings. However, rather than this rendering her artistic engagements a failure, Hawkins contends that it is the *process* of drawing (not the finished product) which is important (2015: 255). As she notes, it matters not whether her drawing resembled the object of her attention – what mattered more was that in drawing it she was engaged in an extended period of carefully observing and making sense of her surroundings. It was this process that generated the geographical knowledge she came to glean about the space and place she worked in. Therefore, do not be put off from creative approaches for fear you lack the skill. Remember that the journey to knowledge is part of the course of working with creative and experimental methods.

Key readings

Hawkins, H. (2015) 'Creative geographic methods: Knowing, representing, intervening. On composing place and page', *Cultural Geographies* 22 (2): 247–68.

Last, A. (2012) 'Experimental geographies', *Geography Compass* 6 (12): 706–24.

Sensory and mobile methods

Tim Edensor's engagement with the ruinous elements of Manchester's cityscape (2005; 2007) demonstrates a creative enfolding between people and place. For a start Edensor contends that ruins are a fertile research ground, offering access to a past not preconditioned through heritage. He also argues that the material landscape of ruins – decaying, broken, uneven and debris-strewn – permits a sensory experience that is missing from the increasingly sanitised city, where embodied engagement is managed (with smooth pavements, moderated soundscapes, and ambient fragrancing). Through investigating the ruin, Edensor notes how people move in ways that are no longer governed by the conventions that usually control how we might orientate the city. Embodied movement is less regulated, more playful. Engagements have a greater curiosity and inquisitiveness. He also argues that the ruin offers a greater sensory experience through the variety of disruptive aesthetics that the eye beholds, the repulsive odours that might be smelled, the amplified silences and sounds, and the material textures which are felt at sites of ruination. In short, Edensor's work is an exemplar of the rich geographical insights we can gain through considering sensory, embodied and mobile engagements with space and place.

In the previous section, we saw that creative and experimental approaches (theatre, drawing, knitting) require a certain embodied engagement where the researcher is closely involved in data collection. The body, in other words, is a central medium in making that data generation possible. Traditionally, the role of the body has been neglected in academic scholarship (Longhurst, 1997: 492). Yet today the body is far from 'abandoned' in the social sciences. In order to understand our engagements with space and place we have to be alert to bodily sensations (Rodaway, 2002) and embodied mobilities (Cresswell, 2006). So let's begin with the former.

Geography has typically been an ocular-centric discipline (Rose, 2003: 213). However, there is now a wide appreciation that the full range of bodily senses are vital for understanding engagements between people and place. Consider the sense of sound, for example. For Atkinson (2007), city spaces are as much defined and made 'knowable' through aural dimensions as they are from their visual features (skyscrapers, transport networks, bustling streets, and so on). Indeed, he notes that our understandings of the urban can be enriched and deepened by interrogating the shifting soundscape of the city which is indicative of metropolitan life (Atkinson, 2007: 1905). Sound, in other words, creates novel knowledge about the city which is impossible to grasp through vision alone. It has been adopted by geographers in all kinds of novel ways to create new geographical knowledge. This provides much inspiration for student projects. Indeed, Pamela Moss's work (1992; 2011), for example, has analysed music as text (see Chapter 7), exploring how the lyrics of Bruce Springsteen have constructed specific representations of American life and landscape at various times throughout his 40-year musical career.

Geographers have also interrogated sound to make sense of social exclusions. The work of Philip Boland (2010) and Anja Kanngieser (2012) has investigated the sonic meanings encapsulated in accent, and how the sounds of our voices are 'imbued with socio-political connotations… (of) geographical background, class, race, nationality, education' (Kanngieser, 2012: 341). Dialect and accent can often be 'placed' – both spatially (to Liverpool, Birmingham, London, Swansea or wherever) and socially as 'moral, political and cultural judgements' are made (Boland, 2010: 4), leading to what David Matless (2005: 758) has called 'sonic exclusions'. Furthermore, geographers have embraced sound in experimental and creative ways (see Butler 2006, 2007; see also the previous section). Toby Butler, for instance, has created a series of 'sound walks' – novel, artistic accompaniments which visitors listen to as they engage with the city. Butler demonstrates how hearing oral histories, stories, music, and so on, together with walking in the city, helps visitors experience place in unexpected and nuanced ways.

And it isn't just sound. Butler's work alerts us to the fact that walking in a city is a multi-sensory experience, physically felt through the body. Touch has provided a novel way of getting to grips (literally) with spatial experience (see Dixon and Straughan, 2010; Paterson, 2009; Wylie, 2005). Here approaches centre on the haptic or felt engagements of the body with place. Through considering the everyday practice of walking, for example, geographers have been able to consider the socialised conventions of rambling (Edensor, 2000); how we immerse ourselves within place by physically feeling the landscape through our feet (Ingold, 2004) and via muscular sensations as lactic acid builds in our calves (Paterson, 2009); and how the body becomes enfolded with landscape through such sensed experience (Wylie, 2005) (Figure 8.1).

But how exactly do we research embodied experience? How do we retain a sense of lived experience and all its embodied feelings when we come to record it as 'data'? Authors such as John Wylie (2005) have demonstrated how sensuous and embodied geographies can be communicated in a way that does not reduce the very feelings such work hopes to understand. Wylie's employment of **autobiography** achieves a deeply personal, critically reflexive account of what it was to walk the South West Coast Path (see also the Graduate Guidance offered in Chapter 1). To read Wylie's account is to be there, sensing the walk with him. This is achieved through a written account that breaks with academic tradition, wandering (much like the walk) from point to point, creatively merging image and text.

Walking alerts us not only to the ways in which geographers are engaging with different forms of embodied knowledge in their research, but also with knowledge that is made in moments of **mobility**. The 'new mobilities paradigm' (see Sheller and Urry, 2006) is concerned with how social worlds are constituted by and with movement at the micro-scale of the body, to the macro-scale of global trade, whilst also considering the range of virtual and imaginary movements that make up the ways in which we are mobile in everyday life (see also Cresswell, 2006; Urry, 2007 and Chapter 2). However, this new way of thinking, it is argued, demands new

Figure 8.1 Doing embodied research: Students make use of mobile methods during fieldwork in Aberystwyth. (Source: Kimberley Peters)

methods (or at least a reinvigoration of current techniques). Law and Urry (2004: 403–4) have noted that 'classic' methods, such as interviewing, focus groups, questionnaires, and so on, are typically static. They take place, *in place*, and as such deal 'poorly' with capturing the range of movements that are so intrinsic to the contemporary world (Law and Urry, 2004: 403–4). It has been posited, therefore, that mobile methods are required for researching elements of mobile life (although see Merriman, 2014, for a critique).

Mobile methods might be defined as those where 'the research subject and the researcher are in motion in the field' (Ricketts Hein et al., 2008: 1267). In other words, rather than trying to make sense of a world of movement retrospectively through interviewing or a survey, mobile methods are those which aim to study movement as it happens. These methods seek to grasp what it is to move by 'being there' or being 'in situ' with movement (Fincham et al., 2010: 6). In other words, if your research project was concerned with urban walking, rather than asking someone how it feels to walk in particular city spaces, you could conduct a 'go along' interview where you walk with your participants (Holton and Riley, 2014). You could also record motion as it happens, later analysing a 'mobile' record of the phenomena you are studying (see the work of Justin Spinney (on cycling, 2011) and Bradley Garrett (on urban exploring, 2011a, 2011b)). A 'guide' to researching mobilities can be found below (Information Box 8.1).

INFORMATION BOX 8.1
USING MOBILE METHODS

When researching mobilities, geographers might embrace the following techniques:

1) Observational research: Researchers can 'look at' people's movements closely, 'their strolling, driving, leaning, running, climbing bodies, bodies lying on the ground... and so on' through direct approaches such as 'shadowing';
2) Active participation: Researchers can involve themselves in the movement they are examining, resulting in a 'co-present immersion' in the mobile landscape under exploration. This might be thought of as a mobile ethnography. Such participation can often be captured using technology such as video cameras;
3) Time-space diaries: Researchers and participants can keep a detailed record of 'what they are doing and where, how they move during those periods and the modes of movement' allowing the researcher to 'plot' movement and its drivers. Diaries can be in written form, or may be photographed or videoed;
4) Virtual research: Researchers can investigate the virtual movements of people through 'analysing texting, websites, multi-user discussion boards, blogs, emails and list-serves'. Such research presents ethical challenges (see the next section) and the extent and direction of movement is not always transparent or easy to follow;
5) Imaginative research: Researchers can use art and design (for example, drawing and painting) to try and capture some of the imagined mobilities of participants, or use gaming technologies to make sense of imagined futures (for example, mapping out ideal city designs);
6) Following: Researchers can follow the journeys of people and things exploring how travel impacts subjects and objects and the decision-making processes that determine movement. Technologies such as GPS (Global Positioning Systems) can be embraced to enhance such approaches.

(Adapted from Büscher and Urry, 2009: 104–8)

Go online! Visit **https://study.sagepub.com/yourhumangeography** if you would like to read more about how to employ sensory and mobile methods in practice. Here you can find additional material outlining how autobiography and videography can help us capture embodied experiences and movements.

Key readings

Büscher, M. and Urry, J. (2009) 'Mobile methods and the empirical', *European Journal of Social Theory* 12 (1): 99–116.

Wylie, J. (2005) 'A single day's walking: Narrating self and landscape on the South West Coast Path', *Transactions of the Institute of British Geographers* 30 (2): 234–47.

Researching online

It might be argued that the internet is intimately connected to topics of embodiment and mobility. On the one hand, geographers have discussed the impact of virtual technologies on the importance of the physical body, questioning whether new technology renders the body obsolete or allows flexibility in the (re)making of identity (see Shields, 1996; Kitchin, 1998). On the other hand scholars have demonstrated the mobilising capacities of virtual technologies to compress time and shrink distance (Massey, 1997). Indeed, in their pivotal work, which has done much to increase our understandings of using online methods, Madge and O'Connor have demonstrated how internet-based research can solve some of the issues that arise in researching groups that lack mobility (2002; 2006).

Indeed, in investigating the role of the internet for new mothers, Madge and O'Connor used online methods to conduct their research with a group that was geographically dispersed. Moreover, given the project sought to understand uses of the internet for new mothers (or e-Mums) it made sense to embrace the internet as a method for accessing this group. Two forms of internet-based tools were employed, an online survey and a group interview. Data revealed that new mothers could use the internet to perform various parenting roles, with the internet itself providing greater freedom as a research tool for articulating this. The online format of surveys and interviews allowed participants to express themselves in a way not possible through face-to-face engagements between the researcher and researched. In turn, the project found that the internet was a source of empowerment for new mums, as well as an empowering medium for the researchers who used it to make sense of such experiences.

We can see from this example that internet-based research has potential. The internet is now pervasive in most of the western world with access enabled on home computers, laptops, tablets and mobile phones, and connection possible through a web of Wi-Fi and wired connections that makes going online possible *virtually* anywhere. Whilst there is still a large 'digital divide' and the web is not available to everyone, everywhere, it remains that the internet is providing a host of new avenues for geographical research. For a start the internet is permitting new kinds of research because it is often more economical 'in terms of time and money' (Bryman, 2004: 470). As Alan Bryman notes, conducting surveys online can radically reduce the direct involvement of the researcher (who can post the survey at the click of a button, or send interview questions via email). This is often crucial for student researchers who have a limited time-frame to complete the study, and often limited resources to conduct the research.

Online research is also broadening the geographic reach of projects, given the capacity of the internet to contact people many miles away, instantaneously (Madge, 2010: 177). Students, therefore, might be more adventurous in the places they are able to research through the internet (see Chapter 5). Finally, internet-based research tools often allow us to collect, and also collate, data very quickly. An online survey, for example, will enable a rapid response to reach the researcher, and often the software used will analyse the data at the click of a button. Likewise, interviews online (where the participant types their response) saves the researcher time in listening back and transcribing the data.

That said, internet research does have its pitfalls and such methods won't be suitable for every project. It is important for researchers not to be seduced by the cost-cutting, time-saving potentials but to think seriously (as Madge and O'Connor did) about whether internet-based approaches are appropriate, and whether they will aid data collection in light of the research question asked. If you decide to use online tools for data collection, it is necessary to consider the points raised by Bryman (2004: 470) in Information Box 8.2.

INFORMATION BOX 8.2
A GUIDE TO USING ONLINE METHODS

The following questions should be posed before you engage with online methods. This will gauge their suitability for your project.

1) Do your participants have access to the internet? For example, populations who have low incomes, or live remotely might be better approached with a drop and collect survey rather than an online one.
2) What is the sample you hope to reach? With internet-based research, the span of your research can be geographically very wide. It can also be harder to pin down an appropriate sample (using the sampling techniques discussed in the questionnaire section of Chapter 7). It is important to consider, then, who exactly you want to reach, rather than 'firing off' your emails and survey to just anyone.
3) How will you actually find and access your participants? It is often a myth that internet research is easier/simpler. It might be quick to send out a survey, for example, but how do you actually find and access suitable participants to whom you might send it? Unless you are using personal networks, this can be tricky and requires researcher consideration in advance, to ensure methods can be deployed successfully. Madge and O'Connor, for example, used an existing forum to access participants. Consider if you can find an entry point, or gatekeeper to the (online) community you seek to address.
4) In what ways will you build trust and rapport? One of the real disadvantages of internet-mediated research is the 'loss of the personal touch' (Bryman, 2004: 270). With online interviews or focus groups it can be difficult to build the rapport vital to establish trust and to encourage participants to 'open up'. This is even the case with technologies such as Skype where you can see participants but eye contact is less direct, and introductory handshakes are impossible. Researchers have to be smart in ensuring they find ways to connect with participants when using online methods.
5) Do you have a back-up plan for low response rates? As Bryman notes, response rates for online research can be low (emails can be more easily ignored than phone calls, survey links can be discarded to the trash folder, and so on). Researchers should always plan to send reminders to participants (Madge, 2010: 176) but should also consider whether additional methods are required if the data yield is low.

(After Bryman, 2004: 470)

However, we are jumping the gun a little. What types of online methods might you actually use in your human geography dissertation? First it is important to acknowledge that there are different types of online methods, which Bryman (2004: 470) classifies as being either **communication-based** or **web-based**. Communication-based methods rely on technologies such as email or chatrooms where a form of communication is established between the researcher and researched. Web-based methods are those where data is collected from 'www.' resources (webpages, blogs, forum sites, news feeds, and so on). Communication-based methods are used for engaging participants (and techniques might include online surveys, interviews or focus groups). Web-based methods are used for data-mining or 'netnographies' (using the internet as a text or object of analysis, or as a means of observing online communities; Kozinets, 2002). Let's consider these in greater detail.

Online surveys, interviews and focus groups

Online surveys are often fast and efficient to disseminate, with results immediately returned to the researcher. They also permit much flexibility in design and in the kinds of questions you might ask (allowing you to insert images or maps). But they are also often very time consuming to prepare (Madge, 2010: 175). Give yourself ample time to design your questionnaire, and, as is the case with paper-based alternatives, conduct a pilot to ensure your questions are comprehensible and flow logically. Madge sets out some key principles for online questionnaire design (see 2010: 175). Notably, she suggests ensuring that your survey has a clear welcome screen (using your university logo), concise instructions that set out how to answer the questionnaire, and a page that outlines the project aims and participant rights (including a checkbox to give consent). She also recommends that surveys are not too long, and that, like paper counterparts, the questions are clearly worded and the font is simple to read.

Interviews and focus groups can be conducted online synchronously or asynchronously. The former means that the researcher and the participant conduct the interview or focus group in 'real-time' (using emailed questions, or group chat software, or, increasingly, Skype/FaceTime technologies). The latter means the respondent replies to the researcher at a later time (perhaps because they are not in a position to reply immediately). For example, this might include 'an interview question posed by the interviewer in an e-mail that is opened and answered by the respondent some time later, perhaps days or weeks later' (Bryman, 2004: 470). When conducting synchronous research, it is vital to ensure you and the participants can work the software competently (sending instructions if necessary) and that your internet connection is reliable. For asynchronous research it is important to note that respondents may provide different answers from those that might have been generated 'on the spot' because they have had greater time to consider and reflect upon the questions posed.

Data mining and 'netnography'

Data mining is where pre-existing content (from websites, blogs, review sites, news feeds, but also social networks such as Twitter and Facebook) is transformed into a source of information for research. As Bryman notes (2004: 467) '[w]ebsites and webpages are potential sources of data in their own right', as 'objects of enquiry'. In other words, a variety of existing web material can be 'mined' for data that is relevant to any particular study, and then analysed in much the same way we might analyse any textual material (see Chapters 7 and 10). Accordingly, it is important to remember that web material – like any text – does not present facts. Rather, web material is produced to convey a certain message, and may in turn be interpreted in a variety of ways depending on the audience (see Rose, 2012). Web material is not innocent and some might be 'misleading and downright incorrect' (Bryman, 2004: 467). Researchers must therefore engage with websites critically. It is also important to employ rigour when searching for sources, remembering that what you find online will be subject to the search engine used, and also the keywords entered. The internet is also a tricky 'field' to research because it is highly fluid – webpages may become inactive or change frequently (Dwyer and Davies, 2010). Researchers should adapt to dealing with a space that is in constant flux by ensuring that data is recorded and saved for future use (for example in a word processing document), along with web address for referencing.

Netnography is a relatively new term developed by Robert Kozinets and is a play on the term 'ethnography', which refers to the method of participating in research communities in order to best understand them (see Kozinets, 2002, and Chapter 7). Working in the field of business marketing, Kozinets demonstrated the usefulness of engaging with communities online to understand consumer behaviour (2002). As Kozinets and others (for example Mkono, 2011) have shown, communities are often played out online, forging new spaces of social interaction (on forum boards, through gaming sites and so on). Researchers can therefore make sense of social worlds by following debates that emerge in online communities. As Mkono's work on tourist culture demonstrates (2011), reading online review comments of alternative eating experiences in Zimbabwe, posted by visitors to the country, allowed a rich insight into travel practices and to the 'othering' of so-called 'exotic' foods. That said, netnographies do raise important ethical questions regarding consent and privacy.

Online research has, as Madge (2010) contends, altered the ethical considerations of researchers. Researchers need to consider how they will ensure the privacy of those participating in projects through communication-based online methods. Researchers have a responsibility to ensure they use secure websites, and, if the data is stored on a computer, that it is accessible only for the researcher. In terms of consent, this should always be sought and this is relatively straightforward in terms of communication-based methods where the researcher can request the participant signs a form (with an electronic signature, for example). More hazy is the issue of consent in regard to practices such as netnography or data mining. Generally speaking, 'lurking' (where the researcher observes online communities and reads blog and forum posts but without the other users being aware of their presence) is not acceptable. Madge (2010) and Mkono (2011) both provide a useful overview of where/when permission to research should be sought, and where/when it is not normally required. For Madge (2010: 181), online communications that are made in private (for example personal email correspondence or posts on a closed forum that requires password access or group approval to join) should not be used without consent. This is because participants will believe their conversations to be confidential to the person or group they are in touch with. They will not have given consent for the wider circulation or use of the information. In contrast, as Mkono notes (2011: 256), posts that are made online in the public domain (for example on sites such as TripAdvisor) require less obligation on the part of the researcher to gain consent. This is because those posting comments upload them in the knowledge they will be available for others to access, read and use. In doing so, they are effectively granting their consent for researchers to use such data in projects. That said, because of the hazy nature of such sources, it is worth checking your own institutional guidance on what measures you will have to employ in respect of consent and permissions for online research. Moreover, students using online methods should think carefully about the ethical dimensions of their project and also recognise that ethical dilemmas may arise as the research progresses – what is called 'ethics in process' (Dwyer and Davies, 2010).

Key readings

Bryman, A. (2004) *Social Research Methods* (second edition). Oxford: Oxford University Press (see particularly Chapter 23).

Madge, C. (2010) 'Internet mediated research', in N. Clifford, S. French and G. Valentine (eds) *Key Methods in Geography* (second edition). London: Sage. pp. 173–88.

Chapter Summary

- To make sense of complex social worlds geographers are employing a range of innovative research techniques. These methods generate forms of data not possible through traditional means of researching. They often require researchers to develop new skills, whilst simultaneously encouraging more accountable relationships with research participants.
- Participatory geographies directly involve participants with the research process, allowing them to guide the lines of enquiry that shape research, as well as taking a central role in data collection. Participatory approaches aim to make research more accountable, balancing out the power relations between the researcher and

researched. Action-orientated geographies likewise produce more equal relationships between researchers and participants as researchers involve themselves in projects as 'activists' to evoke social change.

- Experimental and creative geographies are those that seek to study, and employ, a host of artistic and inventive approaches – dance, drawing, theatre performance, knitting, poetry, music, and so on – in the making of geographical knowledge. Such approaches are not used for the sake of being innovative but rather enable the collection of different kinds of data than would be possible with conventional methods.
- In order to unpack sensory engagements with space and place, geographers have conducted research that has considered sound, smell, taste and touch, often using autobiographical methods to make sense of embodied experiences. Relatedly, geographers have argued that in studying a mobile world, methods are needed to better capture moments that are motionful, fleeting and transient (for example, videographic methods).
- The internet now provides a novel way to collect data in a cost-effective and time-efficient way. Scholars can also access new research sites and less mobile populations due to the geographic reach of the web. Researchers can employ communication-based methods (such as online surveys or interviews) or web-based methods (where the internet is used to engage with online communities or to mine data). Careful ethical consideration regarding confidentiality and consent is necessary when using online methods.

Key readings

Clifford, N., Cope, M., Gillespie, T. and French, S. (eds) (2016) *Key Methods in Geography* (third edition). London: Sage (features new chapters covering novel approaches in geography).

DeLyser, D., Herbert, S., Aikin, S., Crang, M. and McDowell, L. (eds) (2010).*The Sage Handbook of Qualitative Geography*. London: Sage.

9

SELECTING YOUR METHODS: HOW TO MAKE THE RIGHT CHOICES

CHAPTER MAP

- Introducing method selection
- Choosing justified and appropriate methods
- Triangulating your methods
- Mixing your methods
- Locating your methods

Introducing method selection

Commuting. It is probably something we have all done at some point in our lives. When we think of commuting we probably visualise news stories of Londoners negotiating their way to work amidst a tube strike, or residents of big city suburbs crammed on buses or weaving their way through traffic on bikes. In recent years, the geographer David Bissell has conducted cutting-edge research about commuting culture as part of the 'mobilities turn' in the social sciences (see Bissell, 2014; 2016). Bissell identified a yawning gap in the literature concerning this particular type of everyday movement. Commuting, a practice so enfolded in the lives of many – in the Global North and South – had been both overlooked and under-appreciated. Scholars had largely failed to understand those ubiquitous, routinised and habitual movements of people shuttling back and forth, on various modes of transport, between home and work and back again (see also Middleton, 2011).

Consider your own experiences of travel. They are almost always laden with emotional baggage and embodied pressures. You might be leaving behind someone you love, or meeting an old friend. You might be anxious you'll make the connecting flight. You might feel claustrophobic in the busy concourse. You could be swaying with the movement of the train as you stand in a packed carriage or heaving bags that hurt your back.

Returning to Bissell, in order to best understand and respond to his research question (to make sense of the embodied experience of commuting), he had to consider which methods would allow him to gain such insights. Like all geographers, Bissell's techniques were carefully considered to ensure they were feasible (see Chapter 5) and appropriate in view of the question he asked and the kind of information he sought. It is worth noting that Bissell's question was not concerned with numbers of commuters, or volumes of traffic. He was not aiming to make statistical claims. Rather, he wanted to find out how people experienced travel, intimately, through their bodies, their senses. This required qualitative information – data that would be subjective and personal (see Chapter 2). Bissell's work involved long periods of commuting and critical autobiographical reflections on this form of travel, building up a wealth of rich, deep and analytic field notes. It also involved in-depth interviews with commuters around themes such as stress, and the use of novel, experimental approaches, whereby commuters were asked to draw or map their mobilities on paper (see Bissell, 2014; see also Chapter 8).

When reading an account of Bissell's research, the methods he selected make sense in view of his research enquiry. But how easy is it to select the right methods? Surprisingly, method selection is a trickier business than this example would have us believe. Choosing your methods means reflecting on your approach to geography (see Chapter 2) and deciding what techniques are most appropriate in light of the kinds of claims you hope to make. This chapter details the task of choosing your methods, alongside considerations you may need to bear in mind in employing them. To start, the chapter outlines the need to choose methods that are justified and appropriate in respect of your research question, before turning to the technique of triangulation, where a various methods are used to verify data and give your findings greater credibility. Next up, the chapter charts how you might take a mixed method approach of combining qualitative and quantitative data. The chapter ends by exploring how geography matters to the process of employing methods in terms of where your research is located.

Go online! Visit **https://study.sagepub.com/yourhumangeography** for access to Sage journal articles that attend to issues of method selection and use. The articles here focus specifically on how we mix methods, as well as how we take a geographical approach to data collection.

Choosing justified and appropriate methods

Part of the skill of doing your human geography dissertation is displaying an ability to select suitable methods in relation to your research question. At many institutions (check your own institutional guidelines, see Task 1.1, Chapter 1) the marking criteria will allocate a percentage of marks to your choice of methods and their suitability in allowing you gather data relevant to the research problem you outline. Accordingly, getting method selection right is crucial. Yet how do you ensure that you pick techniques that are justified (that is, relevant and warranted) and appropriate (suitable for the task of data collection)? Some key considerations are outlined below.

Select methods relevant to the question

The methods you choose should allow you to gather data that enables you to answer the research question you have set yourself. Let's return to Chapter 4. As noted here, it is important to think seriously about the wording of your question when it is formulated. Your question should reflect accurately, what you want to find out – do you seek to make a firm statistical claim/show a clear pattern, or do you aim to illustrate the messy complexity of the world? Questions, by default, reflect the kind of methods you want to use. It is surprising, however, that methodological mix-ups can often occur when selecting the right tools to do the job. For example, if a student writes a research question that suggests the project seeks to find a definitive answer (yes or no, significant or insignificant, greater or smaller) we would expect to find that quantitative techniques (surveys or questionnaires) are used. If the student was to use qualitative methods, these would be at odds with the question posed. Methods should map onto the question asked, and the question asked should be phrased in such a way that it reflects the kind of method(s) that will be used (conveying a sense that a conclusive claim is sought, or subjective meanings are desired). Accordingly, when you select your methods, ask yourself the following:

- Does your question wording lead towards a definitive answer? If so, you will need to gather quantitative data (employing survey methods or mapping, for example).
- Does your question aim to uncover multiple perspectives and opinions? If so, you will need to generate qualitative data (using ethnographies or interviews, for example).
- Does your question need 'hard' fact *and* subjective meanings? If so, you will need to engage with methods that glean quantitative and qualitative data (see below on 'Mixing methods').
- Does your question seem at odds with the methods you want to use? If so, you will need to adjust your question so that this, and methods you use, are complementary.

Selecting methods within your comfort zone

Your human geography dissertation – as the longest piece of academic work you will produce during your studies – will already, likely, push you outside of a comfort zone (even if you find it to be an enjoyable task!). For some students, however, the idea of employing certain methods might be worrying for any number of reasons. It could be that you have excelled at statistics, but the idea of writing an embodied ethnographic account is a frightening prospect. On the other hand, it might be that you have struggled with GIS but feel confident talking to people about subjective topics. It will often be the case that the type of project you choose will play to your strengths, as well as reflect your way of understanding your world (see Chapter 2). That said, it is worth thinking about the kinds of methods you would be happy to use and working out how you can feasibly engage with these in your project. This might mean adjusting your topic and question a little (see above and Chapter 4), or it could mean tweaking the scope, scale or timeframe of your project (making specific methods you would like to use more suitable). To provide an example, a student of mine was interested in the geographies of women's reproductive rights (for an example see Jackson and Valentine, 2016). Aware that this was a sensitive topic and that recruiting interview participants might be difficult, the student instead chose to refocus the project, using a historical case study which relied on textual and archival data from newspaper records, letters columns, parliamentary debates and documentaries.

Selecting methods you can employ in practice

In Chapter 5 we discussed practical considerations that must be taken into account when designing your human geography dissertation. At the design stage you should think carefully about whether you can feasibly complete your project in practice (can you afford any travel that is necessary; can you access the participants you need to speak with; is your research safe and ethical?). That said, it is also worthwhile to ask if you have the necessary equipment or materials to employ the methods you have selected. If you intend to complete interviewing, do you have a decent voice recorder? Dictaphones are frequently used for capturing interviews. That said, many mobile phones now have voice recording capacity and these are becoming a useful research tool. If you choose to use your phone, test the quality of recordings by completing a mock interview with family and friends. Also be aware of how long you can record for, as some older phones have reduced memory and will only save short recordings. Additionally, make sure you have turned your phone to silent and/or disabled texts and emails so that the recording is not interrupted. If you are completing focus groups, participatory research or mobile methods and want to employ the use of video, make sure the technology is accessible. Do you already have a handheld or small camera at your disposal or can you borrow one? For surveys, do you have the software to build an online survey, or the printing capacity for producing paper ones? Likewise, for GIS, do you have the data available and correct software to construct your maps? For creative or experimental methods, will you need to buy any specialist art materials? (see Hawkins, 2015).

Accordingly, when selecting methods, be aware that methods often require resources. For example, if your research relies on video capture but you have no way of accessing an audio-visual recording device, you will need to think carefully about reframing your study and selecting a different method to gather data. It is often worth enquiring with your institution as to whether equipment (Dictaphones, cameras, video recorders and so on) are available to loan during the period of your study (see Chapter 5). It is also worth investigating what software, datasets and online programmes are openly accessible for you to use. Every institution is different and will likely have a different provision of tools available so it is useful to find out early on in your research planning (see Section I) if the methods you intend to use are predictable.

Triangulating your methods

When selecting the tools to conduct your research, it might be the case that just one method is justified and appropriate. Think, for example, of a project about the operation of piracy in the seventeenth century and how this shaped the production of geographical knowledge (indeed, pirate ships were unlikely spaces of geographical data collection, as shown by Hasty, 2011). With such a historical focus, archival work would be the most suitable method. We can deduce this by recognising that other methods – focus groups, interviews and ethnography – would be impossible. Research of historical narratives and records is the only feasible approach where relevant research subjects are long dead and buried. It is crucial to remember, when selecting your methods, that it is not the number of methods you use that matters. It might be – as illustrated here – that one method will suffice as appropriate and justified in relation to unearthing data relevant to your research question. It can be tempting to think that the 'best' dissertations use every method in the book. However, a good dissertation is one that engages with approaches that are suitable in light of the research question posed. What matters, then, is using the right method or right methods.

That said, it is common for student projects to draw on more than one technique of data collection (see Chapters 7 and 8). This is because more than one method might be appropriate and justified. However, the most common reason for using more than one method is **triangulation**. Triangulation is a process of using multiple research methods to cement the reliability or credibility of your findings (Information Box 9.1, below). Stemming from a quantitative approach, where multiple datasets improve 'confidence in findings' (Bryman, 2004: 275), triangulation is now used widely across social sciences disciplines, such as human geography, in both qualitative and quantitative projects. Whilst for quantitative research, triangulation focuses on cross-referencing different datasets to establish results that are more statistically valid; in qualitative research, triangulation is now also employed to bring together 'different perspectives and sources ... to maximise ... understanding of a research question' (Valentine, 1997: 112). In many respects, then, triangulation is about corroboration. 'It is an interpretive practice in which researchers examine different data or results in relation to one another' (Elwood, 2010: 102).

INFORMATION BOX 9.1
WHAT IS TRIANGULATION?

Triangulation involves the use of multiple methods or datasets in combination to offer greater validity or to gauge divergent perspectives on a given topic area. It is a process of bringing together different forms of data, considering how they **relate** to one another. Multiple, complementary datasets are used together to help improve understanding. The key drivers for engaging in the use of multiple methods are as follows:

To give credibility to the findings of the project

Relying on one dataset or one standpoint alone may limit the ability of the researcher to convincingly answer the research question posed. Using multiple datasets or numerous perspectives will increase the reliability of the claims you make (Bryman, 2004: 273).

To offer a more complete picture in relation to the research question posed

As Bryman notes, triangulation aids a researcher in creating 'a more comprehensive account of the area of enquiry in which he or she is interested', in short, providing a more robust analysis in respect of the research objectives outlined (2006: 106).

To improve the explanation

Triangulation helps to build a more solid account by drawing on a myriad of viewpoints (Creswell, 2013: 201). Multiple datasets or perspectives help in 'explanation building' where improved links can be established, and some data sets can be used to support others in making an argument in relation to the research aims.

To provide a context to the research project

On occasion multiple methods or additional datasets are not used to corroborate a position or strengthen a claim but can be used to foreground or contextualise a project, allowing the other data collected to better speak to the research problem posed (Bryman, 2006: 105).

But what might triangulation look like in practice? Let's take the example of a project concerned with the geographies of lifestyle sports (see Wheaton, 2013) and the gender politics of surfing (Evers, 2009; Waitt, 2008; Olive, 2016). Lifestyle sports (from surfing to skateboarding, to sailing, to parkour) have grabbed the attention of human geographers in recent years. Researchers have been interested in understanding the spatial expressions of such sports (the use of waves, urban architecture and so on to produce exhilarating embodied engagements with place), and the spatial politics of these activities (in light of uneven access due to race, class and/or gender, and the governance and control of these sports in public spaces). In making sense of how gender matters to surfing practice scholars have typically triangulated methods and datasets to produce robust claims. On the one hand, research in this area has relied on interviews with male and female surfers to gain their in-depth insights into the use of surf space and how gender impacts the practice of this sport. Such research has also been cross-validated by conducting textual analyses of popular surf magazines, adverts and films to see how gender is represented. Much scholarship has also drawn upon ethnographic studies that compare and contrast personal embodied practices of surfing (as a man, or woman) with those of the participants interviewed and the texts engaged with. Together, these methods (interviews, textual analysis and ethnography) – all of which are suitable for the research question posed – allow a robust set of data to be configured that has greater validity because that data is collated from a variety of perspectives.

In terms of actually practising triangulation, two approaches are common. In respect of the first, the different methods that are identified as relevant are conducted side-by-side, with each recognised at the outset as vital in building a more 'complete' picture in relation to the research problem. We can see this in the example of surfing practice outlined previously where interviews and textual analysis complemented each other. In respect of the second, triangulation can emerge as you do your research. You might be conducting a project on international migration and translocal identity formation (see Brickell and Datta, 2011). Your research might rely on in-depth, semi-structured interviews with economic migrants in the UK. Whilst this alone would generate rich data, you may feel that the research would be supported and given greater credibility by using additional methods (secondary accounts of migrants in books and blogs, or survey data on the number of UK migrants over time, to provide context). In short, triangulation may develop as your project develops. Accordingly, in light of your human geography dissertation, when selecting methods, it is certainly worth giving some thought to triangulation. Does your project require multiple methods? If so, how might you use those methods together to give your project more credibility, to improve the completeness of your study, to build a stronger explanation, or to provide necessary context?

TRIANGULATION ON THE BORDER
CORDELIA FREEMAN

During my research as a PhD student I found triangulation to be an incredibly useful and frankly unavoidable part of my methodology. I didn't use the technique to cross-validate one aspect of my research with another, but to gradually help me build a clearer picture of my overall research question.

My PhD thesis was a study of violence on the Chile–Peru border from 1925 until 2015. I was therefore trying to understand phenomena that were nearly a century old while at the same time collecting information on how violence exists in the border region at the present day. In this way triangulation became necessary to search for answers to one, wide-reaching question: *What is it like to live in a violent border region?*

The older historical events that I studied could only be understood through historical archives and I spent roughly half of my time on fieldwork visiting different archives to look at official government documents as well as personal diaries and newspapers. For events that were more recent – around forty or fifty years ago – I sought out oral histories. This meant finding research participants who had been alive in the 1960s and 1970s and interviewing them about what they remembered from that time. Of course memories can be lost or change over time, but this offered a fascinating approach.

When studying the present day I conducted interviews with individuals about their current perceptions of violence on the border and I constructed my own archive from online materials (a form of data-mining, so to speak). In other words, I was collating online data that hadn't yet become part of an official historical record. These online resources were often newspaper articles (which often had interesting comment threads below each post) and social media materials. Using social media was challenging because it was so vast and, as such, it was difficult to find exactly what I was looking for. I was lucky enough to find Facebook groups which perfectly aligned to the ideas central to my research question, and so I was able to collect comments (while being careful with anonymity issues and ethics) as well as 'memes' which provided humorous but surprisingly multi-layered dimensions to my work. I also used other types of social media such as Twitter, as well as the comments that are found below YouTube videos.

The combination of these three methods – traditional archives, interviews, and online resources – allowed me to build up a rich picture of life on a violent border over a ninety-year period. I was able to collect fragments of information from these mixed methods that when compiled together gave a much more detailed and nuanced overall picture. Although it's a cliché, for my research at least, triangulation meant that the whole was greater than the sum of its parts.

Mixing your methods

Whilst triangulation involves using more than one method, there is a growing debate in geography concerning the use of so-called mixed methods (see Sui and DeLyser, 2012; DeLyser and Sui, 2013; 2014). Mixed methods research refers not simply to the use of more than one method, but, rather, it is an approach that draws on both quantitative and qualitative techniques and the subsequent datasets that emerge (Information Box 9.2, below). Typically, in human geography, such mixed approaches have been adopted cautiously because at a fundamental level these two approaches are believed to be incompatible. As Elwood explains, quantitative and qualitative approaches 'rely on different underlying assumptions about data collection and analysis, as well as knowledge and knowing' (2010: 96). As we saw in Chapter 2, quantitative research emerges based on an ontological belief (a way of knowing) that understands the world scientifically. This understanding lends itself largely to statistical approaches, where figures and numbers can be extracted from the world, to explain the world (for example, the percentage of a population that is unemployed and the spatial patterning of that unemployment). Qualitative research emerges from an ontological belief that understands the world hermeneutically (as formed by subjective interpretation and meaning). This understanding leads to idiographic approaches, where meanings, desires, opinions and perspectives are sought that are individual and cannot be reduced to patterns or trends. Due to this divergence, the two forms of data collections are traditionally held in opposition (see Bryman, 2006; Elwood, 2010; Sui and DeLyser, 2012). Consequently, many student projects tend to use one approach or the other.

INFORMATION BOX 9.2
METHODOLOGICAL 'MASH-UPS'

Nowadays, many student projects are 'mixing' or 'mashing-up' methods (to use Sui and DeLyser's term, 2012). Mixed methods are not suitable for every project but they can often be used in the following circumstances:

To provide complementarity

Mixed methods can provide approaches that are not starkly in contrast but, rather, complement one another by clarifying 'the results from one method with the results from another' (Green et al., 1989: 259). In other words, mixing methods can help provide a clearer picture in relation to a research problem.

For expansion and enhancement

It might be that although a suitable method is selected for a project, it fails to yield a decent dataset, or provides useful, but ultimately limited perspectives. Mixing methods can deepen or broaden projects, increasing the range of data in relation to the question posed.

To offset any weaknesses in data collection

Quantitative and qualitative approaches each have their limitations (see Chapters 7 and 8). However, 'combining them allows the researcher to offset their weaknesses to draw on the strengths of both' (Bryman, 2006: 106). Whilst mixed methods are not always suitable, for some projects integrating approaches can build a stronger dataset in relation to a research problem.

For diversity in understanding

In bringing together two divergent approaches, projects that mix qualitative and quantitative methods can offer a wide-ranging view on a particular topic – for example offering up generalisable trends alongside personal reflections (see Kwan, 2002; Sporton, 1999; and Winchester, 1999 for good examples).

However, through triangulation (and choice) mixing methods is now not only more frequent, but is also more necessary (see Jackson, 2011). In a series of reports in the journal *Progress in Human Geography* Dydia DeLyser and Daniel Sui urge scholars to 'bury the qualitative–quantitative divide' (Sui and DeLyser, 2012: 111). They note how popular topics in geography demand a mixed approach in order to fully comprehend the complexity of current spatial concerns shaping the discipline. For example, geographical work on the relationship between human life and the environment (so called 'hybrid geographies' – see Whatmore, 2006; see also Clark, 2010; Yusoff, 2013) is increasingly unable to 'do' research that understands the world as *either* scientific or hermeneutic. This is because the natural world and our engagements with it requires an approach that is about hard fact *and* subjective meaning (Sui and DeLyser, 2012: 113). To grasp how we engage with a changing planet we must know the planet scientifically (its technical processes) and how we relate to it (our personal connections).

The mixing of quantitative and qualitative approaches is also common in human geography studies that focus on population dynamics and human spatial behaviours. Kwan and Ding (2008) (alongside others, see Cope and Elwood, 2009; Elwood and Mitchell, 2012) have demonstrated how a quantitative tool of representing data – the map produced through GIS techniques – can also be used for qualitative purposes. In seeking to find out how Muslim women negotiated city spaces after the events of 9/11, Kwan and Ding demonstrate the richness of bridging across quantitative and qualitative approaches for geographical understanding (2008). In a study principally conducted by Kwan, the authors describe how they sought to understand how a post-9/11 increase in hostility and hate crimes targeted at Muslim women altered women's 'daily activities and travel, access to and use of public spaces, as well as perception of the urban environment (especially their perception of safety and potential risk in the urban environment before and after 9/11)' (Kwan and Ding, 2008: 453). Accordingly, the authors used the following approaches that spanned the quantitative–qualitative divide:

> First, an activity diary survey was conducted ... Each activity diary recorded data for all activities that the participant undertook in the survey day, including their starting and ending time, travel mode, street addresses, and purposes (e.g., household responsibilities, recreational or social purposes, etc.). Oral histories were then collected through in-depth interviews from each participant shortly after the activity diary survey. They are the participants' stories about what kind of changes 9/11-induced hate crimes might have brought to their daily activities and trips, and to their perception of safety and risk in the urban environment ... In addition, participants sketched on a map of the study area to indicate the locations they frequent in their daily lives and the areas they consider unsafe before and after 9/11. (Kwan and Ding, 2008: 453)

Consequently, Kwan and Ding then used GIS to integrate surveys, oral histories and sketches to create a qualitative-quantitative representation of spatial experiences in the city.

But how might you use mixed methods yourself? Firstly, we must rewind to an earlier section of this chapter. Before contemplating using mixed methods, you have to ask if this approach is justified and appropriate. Although the idea of 'mashing-up' methods might be appealing, some projects are not suited to such an approach. Let's return to Hasty's work on the world of offshore piracy in the 1600s (2011). There is little need to mix methods to understand how pirates forged geographical knowledge. The information needed could be gleaned from countless archive records charting the voyages of William Dampier and other corsairs of the seas. So, to begin, you should always ask whether mixing methods is necessary or even desirable. If it is, there are no hard or fast rules for how you might mix methods in practice. As Bryman notes, 'there are relatively few guidelines about "how, when and why" different research methods might be combined' (2006: 99; citing Bryman 1988: 155). However, mixed methods can be used in human geography projects in numerous ways.

Simultaneously

This refers to using quantitative and qualitative approaches concurrently. An example of this can be seen in Kwan and Ding's work where oral life histories and GIS mapping are both integrated at the same time. Simultaneous mixed method approaches tend to emerge at the outset of a project where the research question demands a combined approach. Winchester's study of the role of fathers in lone parent families in Australia, for example, demanded both a quantitative approach, using surveys to gauge a broad 'range of information about the characteristics of this group', and also interviews 'which provided astonishing depth on the causes of marital breakdown and post-marital conflict' (1999: 60).

Sequentially

This relates to the use of one kind of approach to data collection (quantitative or qualitative), followed by the other. A sequential use of mixed methods normally emerges as one approach develops to support the findings of the other. For example, transport geography – the study of the spatial organisation of transport systems at local and global scales – has often depended firmly on quantitative methods of modelling and mapping to make sense of how such networks function (see Goetz et al., 2009). Now, qualitative methods are increasingly used to bring 'plurality' to the discussion and to understand how transport systems are experienced (Goetz et al., 2009: 330).

As a matter of priority

In spite of the frequent use of mixed methods, Bryman notes that there is a limit to genuine integration as most scholars will still prioritise one approach over others (2006). Most scholars tend to lean towards one way of understanding people–place relations, and this ultimately shapes the extent to which methodological mash-ups occur. Indeed, returning to transport geography, Goetz et al. remark how, in spite of a shift towards qualitative analysis, 'the quantitative tradition is very strong and will continue to play a vital role in the future of transport geography' (2009: 331).

TASK 9.1

Consider the research question that you have been developing. Ask yourself which of the methods (outlined in Chapters 7 and 8) will allow you to gather data to answer that question. The following prompts may help guide you:

- Does your research question rest on a particular way of seeing the world that seeks to develop generalisable and objective knowledge (requiring quantitative approaches), subjective and specific knowledge (requiring qualitative methods), or is it open to both?
- Could your question be answered using one method or multiple methods? What method(s) would be suitable? Why? Can you also explain the method(s) that would be unsuitable and state why they would not be appropriate? (This helps in rationalising why you have selected the method(s) picked.)

Locating your methods

So far this chapter has considered how you might select, use and combine methods. But as geographers it is also vitally important to think about how space and place shape the methods we use. When we select methods we must think about whether their use is appropriate and justified in the context in which our work takes place. We must also consider how space and place are implicated in the use of our methods (influencing how we use methods). In short, research does not occur outside of a geographical context, and in turn this context shapes how we 'do' research, as well as shaping what we find out (as seen in Chapter 6). It is now widely acknowledged in geographic thinking that space is 'not simply a blank canvas' on which social action occurs (Holton and Riley, 2014: 61). Rather, social, political, economic, environmental and cultural life is enfolded into space rather than space simply being an abstract dimension upon which lives are lived out (see Agnew and Duncan, 1989; Cresswell, 2004). Accordingly, when selecting methods it is useful to situate, place and *locate* these techniques in our studies (see Anderson et al., 2010). In what follows, we track through three ways in which place matters to method selection, the use of research techniques, and the interpretation of data generated.

Space, place and the development of 'critical' method selection

Part of the process of selecting methods requires a critical and careful consideration of the geographical space you are studying. This might mean recognising if there are cultural or political sensitivities that might make some approaches or techniques inappropriate (see Phillips and Johns, 2012: 69). It may also mean considering whether participants will be able to adequately engage in your research (will they have the skills to engage in online research, for example, or use the technologies frequently associated with mobile methods?). It can also mean contemplating the traditions of different places where home-life or work-life routines could prevent you being able to conduct the study due to potential participants not having the time (or even inclination) to engage in the research.

Geographers working in the sub-field of development studies have been particularly helpful in teasing out the ways in which space is part and parcel of the selection of methods and their subsequent use. Development geography (broadly defined) is interested in the ways in which processes – from global scale and national level policies, to grassroots interventions – shape, address and redress inequalities between the Global North and South (see Willis, 2009). Research in this area – which is often concerned with revealing the power dynamics that make places more or less 'developed' – has attempted to 'place' methods in recent years. Raghuram and Madge (2006) have noted how the western researcher (often coming from a position of privilege) typically decides what to study, whom to study and where to study in the Global South. This subjugates research participants from the outset. Development geographers and post-colonial scholars (see Chapter 2) have subsequently sought to select research methods that are watchful of these power dynamics. For example, it is now common to use participatory methods (see Chapter 8) in development research (Raghuram and Madge, 2006: 273). Such approaches allow participants to take a role in the design of research questions and the 'doing' of data collection. This can then often foster a more ethical and equal relationship between the researcher and researched. Other methods may be less suitable for research with vulnerable or marginalised groups located in the Global South because they cement the hierarchy of power that pre-exists (Raghuram and Madge, 2006: 270).

Space, place and enhancing the use of methods

If we acknowledge that research 'takes place' in place, we can consider also the ways in which the setting of our research has the capacity to enhance what we do and how we do it. For example, in some of my own work concerning tourist souvenirs as geographical objects – and their transition from the place where they were purchased to the everyday space of the home – I conducted a series of interviews with people who collected and kept souvenirs in their own houses (see Peters, 2011b; 2014). The benefit of conducting research in the home was that rather than relying on participants to describe where they kept their souvenirs, they could show me. In showing me, this often prompted a discussion of why items were kept where they were (on display, hidden from view or used for everyday purposes). In short, as Holton and Riley have argued, considering where you employ the methods you select and using methods in the spaces and places you study can allow 'research participants to narrate their lived experiences ... and may serve as a prompt or cue for respondent narratives' (2014: 59-60).

Space, place and the shaping of geographical knowledge

As we saw in Chapter 6, who we are as researchers – our positionality – shapes the research we conduct and knowledge we produce. Yet arguably where methods are employed also plays a role in this knowledge production. Where you choose to conduct an interview, focus group or the geographic scale or extent of the datasets you employ, will ultimately shape the findings of your study. Let's take an example from electoral geography (see Johnston and Pattie, 2011 and their respective work). This sub-field of political geography has been traditionally interested in the geographies of electoral systems (the boundaries of constituencies, the process of gerrymandering); and patterns and trends of voting practice as they are mapped spatially. However, in an important paper, John Agnew (1996) argued that electoral geographers were concerned with the spatial distributions of votes, but explained such distributions with 'non-spatial' factors (income, transport, class, and so on) (Agnew, 1996: 129). Agnew contended, however, that space and place were integral factors that shaped knowledge of elections. As he noted, 'space, across a range of geographic scales, figures in the rhetorical strategies of parties, the nesting of influence processes, and the political geographies of electoral choice' (1996: 144). Context, in short, matters to how we know what we know.

Towards a 'polylogic' approach

This leads us to what Anderson, Adey and Bevan (2010) call a 'polylogic' approach to research methodology. Arguing that the 'where' of method has 'received less attention' (Anderson et al., 2010: 590), the authors develop an approach that moves a consideration of appropriate and justified method choice away from simply thinking about who and what is researched, to the role of place and context in shaping research. In short, they argue that 'geographers could operationalize their methodology "as if place mattered"' (Anderson et al., 2010: 590). Such an approach demands that geographers, when they select methods, think seriously about the location where the method is used and how it intersects with their social position, the person or practice they are studying, and the place they are examining (see Figure 9.1). In sum, when selecting methods, it could be useful to employ a polylogic approach and think carefully about where those methods are conducted and how that location shapes the research practice and knowledge that results.

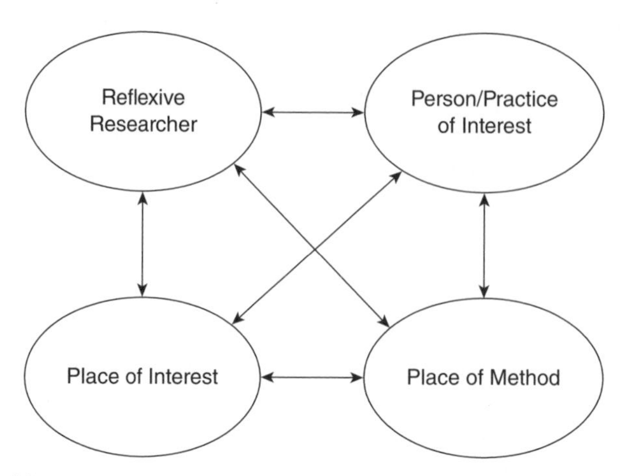

Figure 9.1 A 'polylogic' approach: integrating researcher, place of interest, place of method and person/practice of interest together. (Source: Jon Anderson, 2010)

Go online! Visit **https://study.sagepub.com/yourhumangeography** to read Anderson, Adey and Bevan's article, 'Positioning place: Polylogic approaches to research methodology' in the journal *Qualitative Research* (2010).

Chapter Summary

- There are various methods that human geographers can use in their research projects. However, it is vital to select methods that are justified (that is, relevant and warranted) and appropriate (suitable for the task of data collection). Justified and appropriate methods are those that allow you to collect data relevant to your research question.
- Methods that are appropriate and justified should reflect the research question you have posed. Methods that are appropriate and justified should also reflect your abilities and 'comfort zone', and should be feasible for you to employ in practice.
- Human geography projects may require the use of just one method, or they may need more than one method. This all depends on the question asked. Some projects triangulate methods to cross-validate findings. Triangulation helps to give credibility to results, to enhance the completeness of the study, to improve explanation and to provide necessary context.
- Some projects require mixed methods that combine both qualitative and quantitative types of data collection. Mixed methods can help productively expand projects, can offset the weaknesses of only using a single approach and can provide a diversity in perspective in relation to the research question asked.

- When we select methods, as geographers we must consider how space and place matters to that selection process. Some methods will be unsuitable or undesirable in certain settings. On the other hand, we might want to consider how space and place can enhance our use of methods. A polylogic approach argues that space and place should be central to how geographers use selected methods.

Key readings

Anderson, J., Adey, P. and Bevan, P. (2010) 'Positioning place: Polylogic approaches to research methodology', *Qualitative Research* 10 (5): 589–604.

Bryman, A. (2006) 'Integrating quantitative and qualitative research: How is it done?' *Qualitative Research* 6 (1): 97–113.

Elwood, S. (2010) 'Mixed methods: Thinking, doing, and asking in multiple ways', in D. DeLyser, S. Herbert, S. Aitken, M. Crang, and L. McDowell (eds) *Sage Handbook of Qualitative Geography*. London: Sage. pp. 94–113.

SECTION III

DELIVERING YOUR HUMAN GEOGRAPHY DISSERTATION

10

DEALING WITH DATA: APPROACHING ANALYSIS

CHAPTER MAP

- Bridging data and knowledge
- What is analysis?
- Approaches to analysis
- Analysing different data

Bridging data and knowledge

This chapter starts with the image and idea of a bridge. A bridge, such as the Humber Crossing (Figure 10.1), is an architectural structure that functions to link two places (in this case Yorkshire to the north and Lincolnshire to the south). These areas would otherwise be separated by the estuary where the rivers Ouse and Trent meet. Bridges are a physical means of connecting and bringing together two areas that would otherwise remain geographically apart. Consider other bridges: the Severn Crossing that connects England and Wales, or the Forth Bridge, which unites the city of Edinburgh more directly with Dunfermline. In all of these cases, bridges are able to surpass geographical constraints – bodies of water – to bring landmasses into closer touch. Indeed, bridges not only connect, they create links. They lead from one point to another. Bridges are not only material entities, though. A bridge can be metaphorical – a concept used to bring any two disparate entities into contact. For example, geographers often talk of 'bridging' the divide between different parts of the discipline (qualitative and quantitative research for instance; see Sui and DeLyser, 2012); or bridging over a gap that has emerged between two trains of thought (such as feminist geography and geopolitics; see Hyndman, 2004).

Bridges are inherently geographical. They work to connect spaces and places, both actually and figuratively. Their capacity to link and connect means they are also very useful for helping us to think about the process of data analysis in a research project. In Chapters 7 and 8 we examined the ways in which you go into the field (or stay in the armchair) to collect data. However, it is all very well having data, but how do we arrive at knowledge? How do we translate our findings into an answer that responds to our research question? Data, in whatever shape or form – Census data, participatory maps, photographs, survey results, observations jotted in a notepad – are just raw data until we do something with them. We have to make sense of them and this occurs through engaging with modes of analysis.

Analysis can be thought of as a bridge – a means of forging a connection between data and knowledge. Analysis is what leads data to knowledge, linking together raw information with ideas and answers. This chapter considers the bridging exercise that is vital to your human geography dissertation. It examines how you use your data: how you interpret it, make sense of it and translate it into findings. This chapter should be read alongside Chapter 11 (Writing up). This is because, as will become clear, writing is also an exercise that assists in transforming data into knowledge. Here, however, we begin by dealing with the question of what you *do* with the data you have amassed. To attend to this question, we start

Figure 10.1 The bridge to knowledge: think of analysis as a bridging exercise that builds a connection between raw data and eventual knowledge. (Source: Jennifer Turner)

by exploring what analysis actually is (and why you should endeavour to process the material you have collected). Next, the chapter outlines the key ways we can interpret the data we have (using inductive or deductive reasoning) and how we work with theory to make sense of our findings. Finally, the chapter concludes with a discussion of how to analyse different kinds of data – from interviews, focus groups and ethnographic diaries, to texts, statistics and maps – and what tools are appropriate for answering the research question we have asked.

Go online! Visit **https://study.sagepub.com/yourhumangeography** for access to a series of Sage journal articles that show how different kinds of data (interview scripts, textual material and numerical data) are interpreted.

What is analysis?

Before we can engage in analysing our data we need to know what exactly analysis is. It can be tempting to assume that an answer will simply pop up from the material we have collected. It is also appealing (after all the hard work of data collection) to simply assume a conclusion from a few memorable statistics or interview quotes. We may also feel the urge to just cobble outcomes together quickly in order to form a response. However, just like a bridge – which is carefully constructed and checked to ensure its structural safety and integrity – analysis requires a methodical and considered approach to ensure that (like the bridge) the final result is robust and strong. A good bridge inspires confidence. Likewise, systematic and measured analysis will ensure the marker has assurance that your findings are credible and valid.

Analysis, in short, involves careful examination of evidence to arrive at findings. It is a process that can only occur once you have your data or evidence. For scientists, analysis normally involves separating evidence into smaller parts to understand how something works as a whole. Physical geographers, for example, might take samples of a layer of volcanic ash (called tephra) and by analysing the individual glass shards contained in that layer, they will then bring the results together to make a larger claim about past climate, to help shed light on contemporary climate change (see, for example, Petersen et al. 2016). For human geographers, analysis is no different. It still relies on breaking down your evidence and 'testing' this evidence, sometimes comparing differing datasets, in order to be able to reach a broader conclusion. As the Oxford English Dictionary notes, analysis is a 'detailed examination of elements or structure' leading to a statement of the result of this (OED, 2013). Analysis, then, is about examining – in depth and in detail – the data you have in terms of its component parts (its elements and structure) in order to reach some form of broader conclusion (a statement of results).

Your data analysis will likely involve the following stages:

Evidence

First you must collate your evidence. It is often recommended that human geographers – just like their physical geography counterparts – carefully label their datasets. In other words, make sure your interview recordings have clear file names, paper diaries have times and dates noted down and are stored chronologically, and statistical datasets are easily identifiable (dataset 1, dataset 2, dataset 3, and so on). Once you have neatly compiled your evidence, you are ready to 'work on' it.

Separation

Like other kinds of scientists, the best way to get to grips with our data is to separate it. Once compiled, this should be easy to do. Accordingly, we can then take each interview file, questionnaire survey, focus group conversation, video recording in turn and use a particular analytical tool to dissect it. The tool we use will depend on the data we have amassed (see below).

Interrelation

As well as considering each piece of data separately, we may also want to bring our data together to see how things connect. For example, from your survey data, what is the link between the employment types of the respondents and where they live? Or, from your interview data, what is the relationship between how a participant described a particular event, and the way it was recorded by the media?

Interpretation

Separating out our data, and also re-connecting it, enables us to get close to our evidence. In short, we will have carefully, rigorously and systematically used our skills to unpack that data. From this point we can start to make sense of our materials. At this point we should return to the research question we sought to answer and ask how the patterns, trends, observations and ideas from our data can be interpreted as a response to that question.

Explanation

Once we have sought to bring meaning to our data we have to make that meaning clear. We have to articulate and explain what we have found out. This means clarifying a position or making an argument (see Chapter 11) that connects the data to the answer of our research question. Explanations often reveal the causes or reasoning that have led us to reach a conclusion. It is a detailed discussion, supported with interpreted evidence, of how we have come to the findings we have reached.

Analysis is a slippery term to explain because it is a process best understood by doing it. In the Graduate Guidance box below, student Jonathan Duckett explains how he analysed interview material in his research project about Scottish citizenship and the Commonwealth Games.

MAKING SENSE OF DATA ON YOUTH CITIZENSHIP
JONATHAN DUCKETT

As part of my PhD research I aimed to explore young people's understandings of national identity and citizenship in relation to the Glasgow 2014 Commonwealth Games and the Scottish Independence Referendum. I conducted focus groups during the lead up to the referendum, after the Games had taken place, with 16 and 17 year old first time voters. I also carried out individual semi-structured follow-up interviews six months later. This qualitative study lent itself to an inductive analytic approach as the research aimed to produce geographical knowledge about young people's experiences of these two significant events. Due to the participants being under 18 years old, it was necessary for the project to be reviewed and receive ethical approval by the University Ethics Committee prior to the start of the study.

Analysis began with the transcription of the focus group recordings. While this took a considerable amount of time, the value of such a task to the research project should not be underestimated. I found that transcription required a different type of listening to the skill of listening to participants within the research setting. Replaying the recordings allowed the research encounters to be partially re-envisaged and analysed. Through this process, I became more attuned to conversational dynamics between participants and uncovered previously unheard side-line exchanges within the discussions. Emotional tones, locked within the data, also became more apparent, as young people audibly expressed a deep excitement or fear towards the very real prospect of a future independent Scotland. Typing their words helped to develop a deeper knowledge of the material, bridging the data recorded as sound and the data produced as text. The rhythm of transcription, scattered with bursts of typing and pauses, created space for analytic reflection. These moments often sparked revelatory insights into the data and provided an opportunity for participant references to be further investigated. I noted down initial ideas, which informed the project's wider analysis and the development of questions addressed in the follow-up interviews.

After the focus groups and follow-up interviews were fully transcribed, the task of analysing the significant stack of transcripts appeared daunting. However, I found that systematically working through the transcripts line-by-line and scribbling down codes when they emerged to be an exhilarating thought process. Successive reading triggered new codes and connections, as I continually moved between data at the line level and considered its significance to the wider dataset. In order to make sense of the web of ideas that was generated, I devised analytic codes that were categorised into themes and subthemes. At times, the process was experimental and creative, which saw the development, demise and amalgamation of categories. However, this was not an attempt to make the data 'fit'. Instead, working through the apparent contradictions and divergence of opinion often revealed valuable and unexpected insights into young people's understandings of the nation. Therefore, the process of data analysis examined the relationships between themes, as well as focusing closely on their constituent parts. Assessing the focus groups and interviews, both together and separately, also bridged the temporality of the data. This analytical stage of the research project helped to explain how the views and experiences of young people in my study varied across the passage of these two unique events.

Approaches to analysis

Now we know what analysis is, we need to think more carefully about how we actually do it. Bridges that are built between data and knowledge rely on foundations of reasoning. The bedrock (if you like) of your analysis (or bridge), will depend on the kind of knowledge you want to arrive at. This might be a definitive claim, or a partial understanding. Accordingly, how you analyse your data will depend on the kind of research question you have asked, the sort of data you have amassed and the geographical tradition in which your dissertation sits. There are typically two ways of approaching analysis and that is through **deductive** or **inductive** reasoning. Let's look at these approaches to analysis in more detail.

Deductive reasoning

Some forms of analysis rely on applying a pre-existing idea to our data. In this instance, the bridge between data and knowledge is reached by examining the materials we have collected in relation to a theory. In other words, we may have a theory and we want to see if our data fits with that theory. It is usual to take this approach if we want to consider whether our data supports an accepted theory, or if it challenges it. This kind of reasoning is most commonly used when analysing statistics. This is because deductive reasoning often seeks to test a hypothesis – that is, we analyse data with the desire to reach a conclusion that proves a hypothesis to be correct or incorrect (true or false). Deductive reasoning often helps us to reach a (more or less) definitive answer. As Kneale explains:

> In deduction, a general idea, such as 'more people live in towns' is formulated as a hypothesis that may be tested, as in: 'population densities increase as you move from rural areas to city centres'. Having defined a hypothesis it can then be tested ... Following analysis, the hypothesis will be accepted, rejected or accepted with caveats. (Kneale, 2003: 84)

That said, any idea can be 'tested' and deductive reasoning is not only used by those embracing quantitative methods. Geographers using qualitative techniques might just as likely bring a 'theory' to their work, to see if and how it shapes their case study. For example, a student might bring theories of Marxist inequality to their study, wanting to see whether, in a given example, relationships between sections of society give rise to unbalanced power structures. Although not testing this through a 'hard' hypothesis, and although shying away from any definite claim, the research is led by a theoretical idea.

Inductive reasoning

Other forms of analysis rely purely on the data collected in an attempt to make sense of it. In this instance, the bridge between data and knowledge is reached by drawing directly on the data itself and what it reveals, independent of any pre-conceived ideas or theories. This kind of analysis is common for qualitative researchers who do not seek to prove or disprove a theory (and arrive at a claim of certainty) but rather who aim to uncover the complexity of geographical phenomena. Indeed, inductive approaches to analysis rely on the evidence at hand to make a claim, but that claim cannot be 'absolutely conclusive' (Parsons and Knight, 2005: 47). For example, if the sun is shining we can argue it will be hot – but this argument is not irrefutable. For example, the sun could be out, and it still might be chilly. Accordingly, inductive reasoning leaves room for dispute because results cannot be entirely proved or disproved.

Interestingly, inductive approaches, rather than being led by theory, lead to theory. In analysing the data in such a way, scholars can begin to generate theoretical ideas. For example, as Kitchin and Tate contend, a study on the geographic spread of disease, analysed inductively, would collect data on the spread of disease and 'then construct a theory which explains disease diffusion' (2000: 23). However, whilst we might associate inductive modes of analysis with qualitative studies, just like deductive approaches it is not exclusive in its use. For example, in a recent study of the geographical dispersal of surnames, in an attempt to map a regional geography along these lines, Longley and colleagues (2011) employed inductive reasoning. Collecting data first on surnames and their locations, the authors were then able to map the clusters of surnames, illustrating that '[s]urnames originate, and often remain concentrated, in specific locations' (2011: 506). From this they were able to develop a theory concerning the distribution and frequency of surnames across regions.

Merging approaches

In reality, however, deductive and inductive reasoning often combine when we engage with analysis. Very few of us – qualitative, quantitative or mixed methods researchers – will come to a research project without some ideas or notions that shape the questions we ask and in turn influence the way we separate our data, examine, interpret and explain it. Likewise, most of us will be willing (and open) to the potential of being surprised by our data findings, allowing our conclusions to be led by the statistics, maps, opinions or beliefs we have uncovered. Moreover, as we research, we may engage with one or another type of reasoning. As John Creswell explains, we may start a project with no set idea, allowing the findings to emerge from the data (2013). However, once we have some generalisable ideas, we may wish to deductively 'test' these to see if they 'fit' with an idea we are forging (Creswell, 2013: 186). In light of your own human geography dissertation, it is worthwhile reflecting on how the reasoning you employ can help you create a bridge towards your conclusions.

Analysing different data

Bridge-building between data and knowledge depends not only on the approach we use for analysis. Like selecting methods (see Chapter 9), some techniques of analysis will be unsuitable for making sense of particular datasets. Other techniques will be highly appropriate. It is our job to select the approach and tool that permits us to construct a robust and reliable bridge that moves our data to sound and convincing knowledge. In what remains, this chapter tracks through some key analytic techniques that human geographers use. This is not a comprehensive guide to such tools (indeed, whole books have been dedicated to modes of analysis and I highlight these in the sections to follow). Subsequently, like the summary of methods in Chapters 7 and 8, this chapter introduces the key ways in which evidence is separated, the relationships between data are explored, the means by which interpretations are made and explanations are reached.

Go online! Visit **https://study.sagepub.com/yourhumangeography** to find a web link to the Sage 'MethodSpace' website. Here you will find a set of compiled resources to assist with qualitative and quantitative data collection and analysis.

Analysing words: from interviews, focus groups and ethnographic data

Many human geography dissertations consist of data collated from spoken or written words (from interviews, focus groups, participatory approaches, and ethnographic diaries). This is typically qualitative data. We cannot run statistical tests on the thoughts and opinions related in this data because the meanings revealed are not quantifiable. If your data falls into this category you will probably want to embrace qualitative analysis. But how on earth do we make sense of, interpret, and explain the significance of people's thoughts, feelings, values, opinions and beliefs?

Transcribing

The first stage is to prepare our evidence, ready for analysis. As we saw earlier, it is crucial to organise data so that it can be worked with effectively. Accordingly, if the data is not written down (for example, we have an interview or focus group recording), the first stage is to translate that information to the written word. We can better make sense of a conversation or debate we have recorded if we can return to it time and time again on paper, as a script. This transformation of the spoken word happens through the process of transcribing. But why should you transcribe your data and not just refer to the recording each time? Transcribing enables us to get 'close' to our data, which helps us to understand it better when it comes to interpreting the meanings contained within a conversation or focus-group exchange. As we listen and transcribe words to paper, we can take a note of potentially interesting themes which arise that are useful when we then code for themes more systematically (see below). Transcribing also enables us to take note of moments that might otherwise be lost. We can jot down, for example, pauses, laughter, hesitation, which can all reveal meanings that might relate to our understanding of a topic or issue. That said, transcribing is often described as a painstaking process. It takes a long time to transform a spoken dialogue into words on a page. Most students find that they do get quicker with practice, but it is worth remembering that transcribing is fundamental to successful analysis, so factor in ample time to get this done. Transcriptions can also be inserted into the appendices of your dissertation to demonstrate the rigour of your work to the marker (see Chapter 12).

If your data is already written down (for example in the shape of an ethnographic diary) it is worth consolidating the observations you have made. Field observations are often made in diary form, as jotted notes, scribbled down in the moment when particular noteworthy thoughts, feelings and observations emerged. It is crucial to write up field diaries more fully after events have occurred, adding extended notes. This builds up a wealth of data that is situated both 'at the time' and 'with hindsight' (where things we might not have felt were important when they happened, arise later to become significant to us). If we have conducted our ethnography using a video camera or other device, just as we would interviews or focus groups, we may choose to transcribe or 'write up' what was revealed in these recordings so we can better analyse them.

Coding

Once prepared, we can code our data. Coding is not a process of scanning the page and picking out the most interesting quotes. If we were to do this we would lack a systematic, rigorous and credible approach, and we may miss something crucial in understanding the meanings held within the material. This could then limit our interpretation and explanation. Frustratingly, there is 'no hard and fast rule' to coding (Crang, 2005: 223). That said, coding tends to refer to a flexible 'assignment of interpretive tags to text (or other material) based on categories or themes relevant to the research' (Cope, 2010: 440). In other words, coding is a structured process of allocating themes to our data. It is a means of organising our data by highlighting and grouping information around topics crucial to our research question.

To begin this process we often start with open coding (Crang, 2005). Here we just read over the transcript, carefully, sentence by sentence. At this stage we may mark up or highlight things that stand out. We may also jot our first few thoughts in the margin. This first stage of coding is followed by closer coding. This is where we produce what are called in vivo codes (Cope, 2010: 446). Here we use respondents own words as 'tags', selecting ones that are significant in light of our research question (does the participant mention a particular issue again and again, for example?) (Cope, 2010: 446). We can list these tags in the margin next to the text. Once complete,

Kim: When you buy souvenirs, for other people or things for yourself, where would you buy from?	**In vivo codes**	**Analytic codes**
Leah: Very good question! I am looking at my stuff and it's from all sorts of places! If I am buying something for myself [pause, thinks] actually, I don't think there are any generalisations! I was going to say, as a rule, looking at the jewellery I have, that I try to buy from a local seller cos I would have the feeling I had bought something 'authentic' [laughs!] like someone has made it from Mexico for example and it has little bits of Mexico in it! If I had just got it in a shop it would just be like going in a shop back home. So yeah- arts and crafts markets and stalls where they have blatantly just been down the supermarket and bought it and stuck it on their stall and pretended they made it! [Laughs!] Or yeah – I think I would generally not buy something from a shop that looks like it is for tourists as you know you are not going to buy something original from the country you are just buying something that has been constructed to look local but that no one local would ever buy!	Local seller Authentic Arts and crafts Markets Pretended Original Constructed	Notion that that local is seemingly 'authentic' Arts and crafts tend to be seen as more 'authentic', 'traditional' Awareness that we are duped into thinking local and craft goods are authentic.

Figure 10.2 Developing in vivo and analytic codes: how to make sense of transcribed information. (Source: Kimberley Peters)

we then go back to these codes and develop analytic codes. Analytic codes seek to order the data to link it to broader themes (Cope, 2010: 446). These codes seek to create a bridge between the in vivo code (what the respondent has said) and what that means (an interpretation).

Figure 10.2 provides an example of coding from a project that sought to unpack the meanings associated with souvenir objects (see Peters, 2011b; 2014). It shows how we create a bridge between data and understanding. Although the example provided illustrates only a short extract, we should ensure we code interviews, focus groups or diary material in their entirety. That said, don't expect to always find themes that relate to your research question. You might find you have pages where there is little to code of relevance. On other documents you may amass many codes. Moreover, it can be helpful to code on the computer. Word processing packages will allow you to search, highlight and make notes on a text. It also means you have an electronic copy of your codes and of segments of text relevant to your analysis that you can cut and paste into chapters of your dissertation when you come to write up the project (Chapter 11). There are also more sophisticated coding packages, and if your institution has one of these, it can be worth exploring its use, though specialised training is often required.

All of this said, whilst coding separates out the text, searches for relationships with our research questions, and helps us reach credible interpretations, we must remember that, like any tool of analysis, coding is not perfect. When we code we can decontextualise information. A participant may make one point we deem relevant to an argument we seek to develop, yet their larger point may conflict with this. In such cases we must take care to read data at the scale of individual sentences and broader conversations. Moreover, following Chapter 6, we must consider the ways in which our analysis – our personal selection of codes – shapes the findings we arrive at, engaging with reflexive practice as we go.

Key reading

Cope, M. (2010) 'Coding transcripts and diaries', in N. Clifford, S. French and G. Valentine (eds) *Key Methods in Geography* (second edition). London: Sage. pp. 440–52.

Analysing texts: from landscapes and architecture, to photos and film

It isn't only interviews, focus groups and ethnographic data that can be analysed using coding. Meghan Cope notes that texts – from archives to diaries – can be analysed using a process of coding (2010: 446). Students who are completing 'armchair' work (Aitken, 2005: 233) may also embrace coding as a tool, applying in vivo and analytic codes to archive documents or any secondary sources (a biography, policy report, blog, and so on). However, there is also a further way that we can interpret and make sense of 'texts' and that is through forms of textual analysis.

Somewhat confusingly textual analysis is also a method – a means of collecting data as well as making sense of it (see Chapter 7). Indeed, for much qualitative data, the process of interpretation and explanation folds more readily into the phase of data collection than is often the case for quantitative material. This is because numbers and figures (as we will see below) need to be processed before we can begin to make links to our research question. With qualitative data collection, answers may begin to readily present themselves as we gather materials (something someone said or a theme cropping up repeatedly in archive records). As we saw in Chapter 7, textual analysis is a qualitative approach to data collection and interpretation. If you are dealing with textual material – a film, documentary, government report, tourist brochure, photograph (the list could go on) – this mode of analysis will probably be beneficial.

Analysing the text, the producer and the audience

Gillian Rose (2012: 19) recommends a tri-part model for unpacking (separating out the text) so you can then make larger claims about the power dynamics that shape the text in question. She states that we should consider the site of **the text itself** (what the text itself contains and reveals), **the site of production** (who has made the text and why) **and the site of audiencing** (who is receiving it and how they interpret it). When analysing any text, geographers should look to carefully consider each in turn.

Starting with the text itself, it is common to examine and make a note of the contents and aesthetics displayed. This means considering the composition, colour or fonts used in the piece (What text is in bold? Where is the eye led? How are objects arranged?). We can then seek to ask why the contents and aesthetics are as they are. What do they reveal about how the text communicates its meaning? Next we may want to examine the producer of the text, asking who made it and why. Texts are never innocent, and they do not depict the world as it is. Rather they are always shaped by their authors. Geographers should try and ask 'what message does the author want convey?' To make sense of this message, we must research who has created a text and link this to the context and aesthetics. Can we ascertain a reason that the text contains or connotes what it does? Finally we should appreciate that texts have multiple meanings. Whilst a producer will intend for us to read a text in a particular way, alternative readings are possible. We should ask if other interpretations of a text are possible. Hall (1980) has argued that audiences can 'accept' texts as the 'truth'; 'resist' their dominant meanings; or they can 'negotiate' what is conveyed (in other words, be aware of the politics of a text, but enjoy it nonetheless). What various ways could the text you are using be interpreted as? Is more than one meaning possible?

Reading for signs

We can also analyse a text by breaking it down and investigating the signs that are used to create meaning. Signs can be split into two components: the signifier and the signified. All texts contain words or images that are signifiers and point towards some kind of message we are supposed to receive as consumers of the text. The relationship between a sign and what it means only makes sense because we are taught to associate signs with particular meanings. We might, for example, know that graffiti and broken windows in a photograph of a city are signs that signify a place is run-down, crime-ridden or poor (see Herbert and Brown, 2006). Confusingly, though, some images might have multiple codes. This is known as polysemy. So we need to decipher which one is right at any given time. The colour red, for example, could be a sign used to signify either danger or love. We need to re-code the colour depending on what context we read it in. Moreover, we should also consider whether our reading of a particular sign is shaped by our knowledge of other texts. As Pamela Shurmer-Smith notes, 'texts are, themselves, constructed not just out of unmediated experience but also in light of other texts' (Shurmer-Smith, 2002: 128). We should ask if the meaning a text holds is based on intertextuality – our knowledge of other texts. When analysing texts, we need to ask ourselves what signs are used and what those signs signify – as well as how those signs relate to context. In so doing we can begin make a bridge between a text and what it means.

Unpacking discourse

Texts can also be understood if we break them down to reveal the workings of power that often underpin them. This is known as discourse analysis. A discourse refers to a constructed idea that is repeated so often it appears to be natural or inherent (see Foucault, 2013). Discourses are powerful because they often establish norms about society. When analysing a text we should ask if there is a hidden discourse of power shaping the ideas that dominate in that text. Take for example Jason Dittmer's work that has focused on the comic book series 'Captain America' (see Dittmer, 2012). On the one hand, Dittmer shows us that a text such as a comic book can be used by geographers to explore space, place and politics. On the other hand, his work demonstrates how a comic works as an influential text that contains geopolitical discourses about the American nation-state. As Dittmer notes, underlying both the character of Captain America, and the narratives produced, is a powerful patriotic message (Dittmer, 2005). Dittmer notes how characters in the comic who mock Captain America (his uniform in particular) are held to be villains who disrespect patriotic ideas (2005: 257). This example demonstrates the importance of dissecting a text, reading it closely and interpreting its meanings. This kind of analysis requires us to deconstruct what discourse is at work, so we can see past it and understand the ways it shapes the meaning conveyed through the text.

Key reading

Rose, G. (2012) *Visual Methodologies: An Introduction to Researching with Visual Materials* (third edition). London: Sage.

Analysing relationships: from statistics and maps to patterns and trends

Just as texts can be analysed using coding, as well as searching for their signs and discourses, statistics and maps can be dissected as texts. For example, some research projects may seek to make sense of the ways in which sets of data are collated and authored. Census data, for example, is not reflective of a population 'as it is' because it will exclude certain groups of people from the count (notably migrants or the homeless) (see White, 2010). Likewise, we might be interested in a series of maps, questioning how their content and production shapes geographical ideas. Maps – global, national, regional or local – often do not convey the world as it is, but represent a way of knowing the world visually. The position of Europe in the centre of most world maps is no coincidence. It represents an idea – or discourse – that Europe is at the 'centre' of the world. However, whilst we could explore statistics and maps in this way (if our question demands it), more likely, we will want to use numerical or mapped data to describe geographical phenomena, to uncover relationships between two or more socio-spatial variables, or to ascertain the probability of a particular occurrence.

Describing

Dorling urges that we should only use statistics when they are useful to us (2010: 374). Numbers, by their very nature, are 'dull as ditch water' unless they can be used to tell us something interesting (Dorling, 2010: 324). In your human geography dissertation, if you are using numbers, they should be helpful in response to your research question. One way in which numbers (whether you have collected them, or they have been sourced from an existing survey, for example) can be usefully employed is in a descriptive capacity. Analysing a dataset to reveal its characteristics is often a key mode of making sense of what the data has to say. Descriptive statistics usually refer to a process of revealing the 'shape' of the data: its mean (the average value in the dataset); its median (the centre value in the range); and its mode (the most frequent value that occurs across the dataset); as well as the spread of the data, quantified by the range of the dataset (difference between the highest and lowest values in a dataset) or its standard deviation (a measure of the spread of the data). That said, we can only apply some of these descriptions to some types of data (see Information Box 10.1, below). Moreover, whilst useful, descriptive statistics cannot explain relationships because they typically refer to single sets of data. As Dorling suggests, it might be interesting to note that 'X was the average number of residents who supported the recycling policy' but 'so what'? (Dorling, 2010: 375). We can only interpret what this might mean if we compare that figure to another.

INFORMATION BOX 10.1
TYPES OF NUMERICAL DATA

When geographers engage with pre-existing data or collect data themselves, that data will have different qualities. These qualities will determine the way the data can be used in analysis. The following are the main categories of data that are collected and used by human geographers:

Nominal data has no numerical value. It can be thought of as categorised data where information is listed by type. Often a label may be assigned to categories for the purpose of analysis. A good example of this might be *gender* where categories 'male', 'female' and 'prefer not to say' are assigned labels '1', '2' and '3'. Other examples of nominal data might include place categories, such as England, Wales, Scotland and Northern Ireland. This kind of data can take any non-numeric form that can be counted but not ordered.

Ordinal data is also non-numeric but data that can be ordered and ranked. In other words, ordinal data is arranged in a rating scale where some values are higher than others. Although we can grade data, the value between the different data on the scale will be unequal and/or unknown. A good example of this is data that reveals if respondents are 'very satisfied', 'satisfied', 'neither satisfied or unsatisfied', 'unsatisfied' or 'very unsatisfied'.

Interval data is numeric, where each figure collected can be positioned in a scale. Data is continuous in nature and the intervals between the data are arbitrary but fixed and equal. With interval data, there is no zero. A good example is temperature. Temperatures exist on a scale with higher and lower degrees possible. There is an equal measure between each figure (1°, 2°, 3°), and there is no 'true' zero because temperature can be below zero (for example -12°, -10°, -8°). With interval data we can therefore gauge the meaningful difference between the variables.

Ratio or scale data is also numeric data, also exists in scalar form, and is also continuous in nature. However, unlike interval data it has a clear zero. This is often the most common kind of data used. A good example is height or weight (which proceeds in equal, spaced, and continuous measurements, but there is a zero – there cannot be *no* height or weight). Another example might be a percentage mark gained in an exam where the lowest mark is 0% (see Kitchin and Tate, 2000: 47).

Correlating

The most common way of attending to the meanings held within your data is to 'identify and quantify relationships between variables' (Walliman, 2011: 213). This means comparing datasets with each other to see how they correlate. It also means ascertaining the strength or significance of the relationship between those variables. A variety of tests (see Kitchin and Tate, 2000 or Rogerson, 2014) can be employed to correlate datasets. The type of statistical test we use to see how significant a relationship is depends on the kind of data we use. Parametric tests (such as the Pearson correlation coefficient 'r') can only be used on data that we consider to be 'normally distributed' (one example may refer to the characteristics of a population where there is a 'central tendency' and an equal spread in the distribution of that data either side of the mean - typically visualised as a 'bell-shaped' curve, see Figure 10.3). These tests are used on ratio and interval data.

Non-parametric tests (such as Spearman's rank or Chi-Squared) can rank data or test the frequency of relationships and these apply to data that has no known parameter (ordinal or nominal data). That said, with any test that correlates data and searches for the relationship between variables, the answer should be taken in context – we can only have a degree of confidence in the result and the strength of the relationship (or rank or frequency) based on the integrity of the data we have amassed (Walliman, 2011: 213).

Predicting

Another useful tool for making sense of numeric data is to use it for inferential statistics – that is to make predictions or test probabilities. Such analytic tests can 'infer' or reach a conclusion from the evidence available. This kind of analysis is used when we want to reach an answer that is beyond the data itself. As Dorling notes (2010: 382), it is of interest to geographers to ask questions about the 'likelihood' of something happening. For example, if the data reveals that 'X' number of people died in a certain area, is it likely that mortality rate will continue? Or, if data over a given time period shows that 'X' number of wildfires occurred in the study area, how many more are likely in the same time period moving forward? In order to analyse data in such a way we need to have a set, representative sample of data to work with. From that sample, predictions can be made 'based on the qualities of a sample' (Walliman, 2011: 213). The kind of analytic 'test' you will use for inferential statistics (whether a t-test, Pearson's or confidence limit, and so on) will depend on the kind of data you have (see Information Box 10.1 above and Rogerson, 2014).

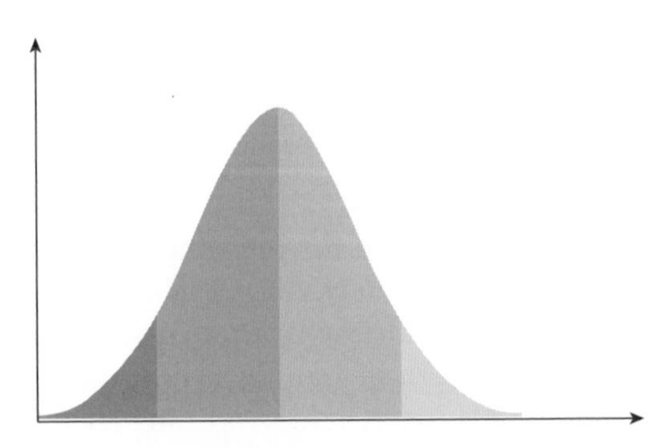

Figure 10.3 A normal distribution: the bell curve. (Source David Remahl (Own work) [Public domain], via Wikimedia Commons)

Mapping

Another way of making sense of statistical data is to map it. This can help us to see, visually, the relationships between society and space, as well as providing a means of then interpreting those relationships because we can observe networks, linkages and connections and suggest reasons for these. We can also produce multiple maps, or create overlays that allow us to compare the same area and same variables (movements, clusters of services, population characteristics) over differing time periods. However, like with all modes of bridging data and knowledge we must ask if a map can be produced from the data we have collected and if the creation of a map (or maps) will get us closer to reaching an answer to our research problem. If it will, it is vital to understand 'how maps work', producing a high-quality illustration (Perkins, 2010: 361 and Information Box 10.2).

INFORMATION BOX 10.2
A GUIDE TO GOOD MAPPING

Attention to the following attributes ensures a high-quality map is produced:

- Maps should convey spatial properties in a consistent way. For example, a map should use a scale (and overlapping maps should all use the same scale for comparison). The scale should be clearly stated and there should be an arrow indicating 'north'.
- The scale should be decided based on the level of detail needed in the map. The larger the scale the more detail can be contained. With a smaller scale, less detail can be mapped. Choose an appropriate scale for your data.
- Objects or symbols on the map may not be to scale (so they can be read) but they should be in a consistent size and style (e.g. settlements, roads or points of interest).
- Often colours, textures, points, lines, objects and symbols can be used to graphically explain variables. These should be referred to in a key that explains their significance.
- Ensure the map you produce is high-quality. This means using appropriate mapping software to produce a clear, balanced and uncrowded map. Choose colours carefully so that objects and symbols are contrasted with the background of the map, and select fonts that make certain textual information legible.

(Taken from Perkins, 2010: 362–6)

Key reading

Rogerson, P. (2014) *Statistical Methods for Geographers: A Student's Guide* (fourth edition). London: Sage.

TASK 10.1

At the end of this chapter it is useful to consider how you might begin to build your own bridge towards geographical knowledge. To do so, reflect upon the following questions. These will help you select the appropriate approach and tools to conduct a detailed examination and interpretation of your data. This, in turn, can help lead towards strong, credible explanations.

- What is the kind of question you have asked in your human geography dissertation? (Does it suggest you will arrive at definitive claims or partial understandings, or both?)
- What is the kind of answer you want to arrive at? (Do you want to be able to provide a more or less certain response, or do you want to be able to reveal multiple, complex responses that cannot be generalised?)
- What kind of data have you collected? (What approach and which tools are most appropriate for making sense of that data? Do those approaches and tools link with the kind of question you've asked and the sort of knowledge you want to uncover?)

Chapter Summary

- Analysis can be thought of as a bridge-building exercise that connects data to findings. Any bridge must be strong and robust for the data to convincingly lead to a final destination – geographical knowledge. The foundation of any analysis must be the research question asked. Analysis that is not conducted in respect of the research problem posed is fruitless.

- Analysis relies on evidence from which knowledge can be claimed. To analyse, students must organise their evidence, separate it out to consider each component carefully, examine whether there are relationships between the different data collected, and finally interpret what that data reveals, arriving at an explanation.

- Projects will adopt an approach to analysis that is deductive or inductive (or, indeed, which draws on both). The former refers to analysing your data in respect of an existing theory or idea. The latter refers to a process of being led by the data, developing conclusions from the bottom up.

- Depending on the kind of question you have asked, the sorts of data you have collected (qualitative, quantitative or both) and the type of knowledge you want to arrive at (certain claim or partial understanding), you will want to use the appropriate tool to dissect the information you have collected. This tool might be coding for themes, analysing texts for discourse, or running statistical tests.

Key readings

Clifford, N., French, S. and Valentine, G. (eds) (2010) *Key Methods in Geography* (second edition). London: Sage (see Chapters 10, 22, 23, 27, 29 and 30).

DeLyser, D., Herbert, S., Aitken, S., Crang, M. and McDowell, L. (eds) (2010) *Sage Handbook of Qualitative Geography*. London: Sage (see Chapters 12, 13, 14, 15, 16 and 20).

Fotheringham, S., Brunsdon, C. and Charlton, M. (2005). *Quantitative Geography: Perspectives on Spatial Data Analysis*. London: Sage.

11

WRITING UP: WHERE TO START AND HOW TO FINISH

CHAPTER MAP

- 'Earth-writing': Getting started
- What to write
- When to write
- How to write
- The 'where' of writing

'Earth-writing': Getting started

As we know from Chapter 1, the word 'geography' refers to a process of 'earth-writing' (Barnes and Duncan, 1992). As an earth-writer, the role of the geographer is to communicate knowledge about the world we live in, through the written word. Accordingly, writing up your human geography dissertation is a fundamental stage in becoming a geographer yourself. Yet writing is very rarely plain sailing. Geographers are trained to think spatially, and we are taught skills and methods that enable us to interpret the world through such a lens. We aren't typically trained to be professional writers. Yet all disciplines require students to write. On the one hand, writing is a way in which we make sense of what we have found out through our analysis (DeLyser, 2010: 433). We can only know what we have found out by trying to articulate it. On the other hand, once we have worked through our ideas, writing is an act of conveying what we know and committing the knowledge that we have produced to paper. Writing therefore cements knowledge – it brings what we know into being.

In a classic paper, the humanistic geographer Yi-Fu Tuan explains the importance of language to making sense of spatial phenomena (1991). For Tuan, geographical knowledge can only come into being through writing. Words that describe geographical worlds come to forge and form those worlds (Tuan, 1991: 692). Accordingly, if language fails us, communication fails. In a dissertation project, if communication fails, the marker will not fully comprehend the work you have completed and the knowledge you have uncovered. Therefore, it is important that adequate time is invested in writing up your research in order to carefully convey what you have found out. Subsequently, this chapter urges you to take seriously the role of 'earth-writer'. Think of yourself not only as a geographer, out in the field collecting data, but as a professional writer who also has a role in communicating ideas to paper (DeLyser, 2010).

In what follows this chapter offers some reflections on writing up your human geography dissertation. The chapter begins by first examining what to write in completing a dissertation. This is the 'nuts and bolts' section that describes the different components that typically constitute an extended research project and how you might plan for these. This is followed by a consideration of when to write. Here, we attend to writing as soon as you can, writing to deadlines and leaving time to draft and re-draft. The chapter then provides some insights into how to write. This section considers the business of constructing an argument, crafting your writing and, lastly, dealing with writers block. Finally the chapter (in true geographical fashion) explores the 'where' of writing, tracking through the best places to work as you write up your research.

What to write

As we know, a dissertation is an extended piece of work (often between 10,000-15,000 words, though some may be shorter or longer). When contemplating what to write, it is often helpful to think not of the 'whole' but to break the dissertation down into chapters to make writing manageable. Each chapter serves a purpose and will build towards the overarching 'tale' of your research (DeLyser, 2010).

The key components

Below we consider the 'typical' chapters that constitute a dissertation and what each one usually conveys. Whilst this list appears rigid, take note of the ability to tailor these chapters to your own project.

Introduction

This is the opening chapter of your dissertation. A good introduction should set the scene for the project. It should provide a clear statement that encapsulates the focus of your study and a sentence or two that explains a rationale for the work you have produced. It should also provide a sense of the context for the work, i.e., does it sit within a particular geographical tradition (Marxism, feminism, post-structural geography)?; what major literatures have inspired it?; and what methods have you used (qualitative, quantitative or both)? The introduction is also typically the place where you would introduce any case studies. A good introduction should therefore provide a taster or an outline of the project to grab the reader's attention. Crucially, it should provide key information but not be too detailed (as greater discussion will follow). Typically a clear introduction will also provide a paragraph that signposts the dissertation to follow (see Information Box 11.1).

INFORMATION BOX 11.1
INDICATING THE WAY AHEAD

Geographers are traditionally credited with being good navigators. Even if this assumption is outdated, there is some merit to considering the role of the geographer as someone who can pinpoint where things are, the routes between things, as well as identifying where paths might lead. Arguably, geographers have all the necessary skills to be excellent 'signposters' when they write.

As an extended piece of work, it is vital that your dissertation is easy to follow. Over the course of so many pages, it is easy for the reader – but also for you as a writer – to get lost. Signposting is an effective way to help guide both you and the marker through your text. Simple statements such as 'in the next chapter, I will discuss this …' can help to link sections together and ensure an argument (see 'How to Write' below) has 'flow' and coherency.

Indeed, it can be helpful to add short signposting sentences at the end of chapters, or between sections of a chapter (especially if you are changing tack, or presenting a new argument). Signposts can be thought of as points on a map that direct the reader towards the next point. As you write, ask yourself when and where it would be beneficial in the text, to help the reader place your argument.

Literature review

A literature review is often situated near the beginning of a dissertation. The purpose of this chapter is to review the work that contextualises your own study. As we know from Chapter 4 it is vital to read around your topic. A literature review should demonstrate to the marker that you are aware of the central debates and more specific work that provide the background for your study (Walliman, 2011: 157). Importantly, a literature review should be integrative. In other words it should not discuss each piece of literature relevant to your project in a list-like fashion. Instead, it should attempt to group literature together around the major themes that underlie your work. Moreover, a good literature review will – through surveying previous work – build towards a statement of what your project can add or contribute to these debates.

Methods

All dissertations require a chapter that explores what you did – the methods you employed and how you used them. Methods chapters have a reputation for being notoriously dull. They needn't be. A good methods chapter will discuss the techniques you used (drawing on the literature to bolster what you say), but it will discuss those techniques in relation to your project. Methods chapters have a habit of describing the approaches used in very general terms (e.g. 'I conducted an interview. An interview is a conversation with a purpose'). Your methods chapter should say exactly what *you* did (e.g. for interviewing it should cover who you spoke to; why; where you conducted the interview; what when well and what didn't go so well; what factors shaped the conversation, and so on). Accordingly, a good methods chapter doesn't simply describe what you did, but critically reflects on the process to show you understand the strengths and weaknesses of your data collection. Indeed, a methods chapter can describe the limitations of your approach as well as the achievements (number of interviews, range of datasets, length of ethnographic study, and so on). It should also document the modes of analysis you used to interpret your data. This shows the marker you can appreciate the factors that have shaped the knowledge you have produced in later chapters.

Results/Analysis/Discussion

This is the one of the most crucial sections of a dissertation because it is where you convey the findings that you have researched and explain their significance. However, it is also where most dissertation projects differ in their structure and presentation. It is vital to think carefully about the kind of dissertation you are constructing and the traditions of writing that are usual in your sub-field of human geography. For example, quantitative projects that rely on statistics, tables, graphs or maps normally present the data in a results chapter and then follow this with an analysis/discussion chapter. Comparatively, qualitative projects which use interview data or ethnographic diary excerpts normally combine results, analysis and discussion together into a longer chapter. Some dissertations break their results and analysis section into two or three separate chapters, each attending to a specific theme/argument (see 'Mapping it out' below). In spite of this variation, what is essential is that you **show your data** (DeLyser, 2010: 430). This means using evidence collected (from statistics to interview quotes) to support your claims. Particularly in qualitative dissertations (where ethnography or interviews or experimental methods are used), students can fall into the trap of alluding to data rather than actually using it. Be sure to weave your data directly into the discussion to illustrate the work you have done (see Chapter 10).

Conclusion

Whilst an introduction sets the scene, a conclusion should tie things together. A good conclusion has three qualities. First, it should summarise your project and clearly state what it has achieved. This shouldn't repeat what has been written previously, but rather should solidify your claims in relation to the research question/problem you set out to investigate. Second, it should attempt to explain 'the broader implications of the work in question' for geographers (DeLyser, 2010: 431). In other words, you might suggest that 'geographers studying A should pay attention to B and C' (DeLyser, 2010: 431). This might feel like quite a bold commitment, but it also helps in thinking through the contribution your work can make to others. Finally then, the concluding chapter might raise the potential for further studies in the area you have examined. Whilst you should avoid 'dropping in entire new ideas' to the dissertation at this stage (DeLyser, 2010: 431), you might propose follow-on studies that develop from your work (for example, whether it would be helpful to explore the same themes in another context or with a different dataset). This takes us back to the wine glass/egg timer analogy in Chapter 4, where a good dissertation often begins broadly, homes in on a topic, before expanding again at the end.

Task 11.1 can help you to think about the kinds of dissertations other students have produced. Moreover, although you have ultimate responsibility for your human geography dissertation, it is always beneficial to discuss your proposed structure with your supervisor and to follow their advice on how to piece together your work. However, before we can hope to write the chapters outlined (in whatever shape or form suitable), first we must plan what to write.

TASK 11.1

It is useful, at the outset, to glean some understanding of what a finished dissertation looks like and the components that come to make up a final report. Most institutions will hold example copies of dissertation projects completed by previous cohorts of students. Find out if your department or school makes these available to students. Look at examples of dissertations that use the same techniques or methods as yours. Reflect upon the following:

- How have they structured the dissertation?
- Does the overall structure seem logical?
- How has each chapter been structured?
- Does the argument flow well?
- What are the strengths of how the dissertation is laid out?
- Can you see any weaknesses?
- What ideas might you bring to your own dissertation from looking at these examples?

Mapping it out

Now we know the core components that comprise a dissertation, mapping out – or planning – your dissertation might seem an unnecessary task. We know, for example, that an introduction leads to a literature review, leads to a methods chapter and so on. However, planning is essential to writing success. Whilst the shape of a project in its entirety is normally quite straightforward, it is the internal organisation and structure of each chapter that often requires more thought. Each chapter of a dissertation is like a mini-essay, but one that must connect to the chapter next to it, in order to combine to create that larger 'whole'. Having an idea of what you will write in each chapter, and mapping out a plan to this effect can be a useful task when you reach the stage of writing up. It can help you to think about the argument and flow of each individual chapter, and how each individual chapter contributes to the argument and flow of the dissertation as a whole.

During your undergraduate studies you will have likely composed plans for other assignments and will have hopefully found a way that works for you (see Kneale (2003) for good advice on this topic). Commonly, when constructing a plan for a regular essay, you will consider the sections of text you will need to write in order to answer the question that has been set. Your essay will normally be flanked by an introduction and a conclusion. Dissertation chapters are no different and we can think about planning them in much the same way. Each one will have an introduction that states the intentions of the chapter, before the main argument, is conveyed. A conclusion typically rounds things up. The hardest task in planning an essay, and indeed a dissertation chapter, is working out what you put in the main body paragraphs. In order to develop chapter plans, first think about the point of each chapter. What is each chapter trying to achieve? The point of a literature review, for example, is to outline the contextual ideas that foreground the project and from which the project emerges. To plan a literature review you would need to consider how to group the different literatures you wanted to explore in your review to make the case for how your research question has developed. To do so you might use a mind map and scribble down all the different bodies of reading that are central to your project and then work out a logical order for conveying them (does it make sense for certain readings to follow others?). Alternatively you might want to list your topics and populate ideas under headings, building a more comprehensive plan (see Walliman, 2011: 317). Try to think of headings and subheadings to split up your text (Parsons and Knight, 2005: 117). The same logic applies to other chapters too. Figure 11.1 provides an example of how we might plan our writing, showing a mind map developed when writing this book.

That said, results, analysis and discussion chapters are often the trickiest to plan. This is because (as we saw earlier) their point will be individual to the study you have completed. It is here that you present your findings, make interpretations and respond to your research question. An effective way to plan these chapters is to think about the data you have (from your analysis, see Chapter 10) and the argument you want to make in respect of the research problem you posed (more on this to follow). You may find that your data naturally divides into key points that will each form a section of the chapter. You may find you need to decide how best to order your data and interpretation to make particular claims. You might choose to structure your discussion around the aims and objectives you have set. What makes the results, analysis and discussion chapters difficult is that there is no set or standard structure that can be used as a guide. Moreover, it is often the case that more than one structure is feasible. Take this chapter for example. It would

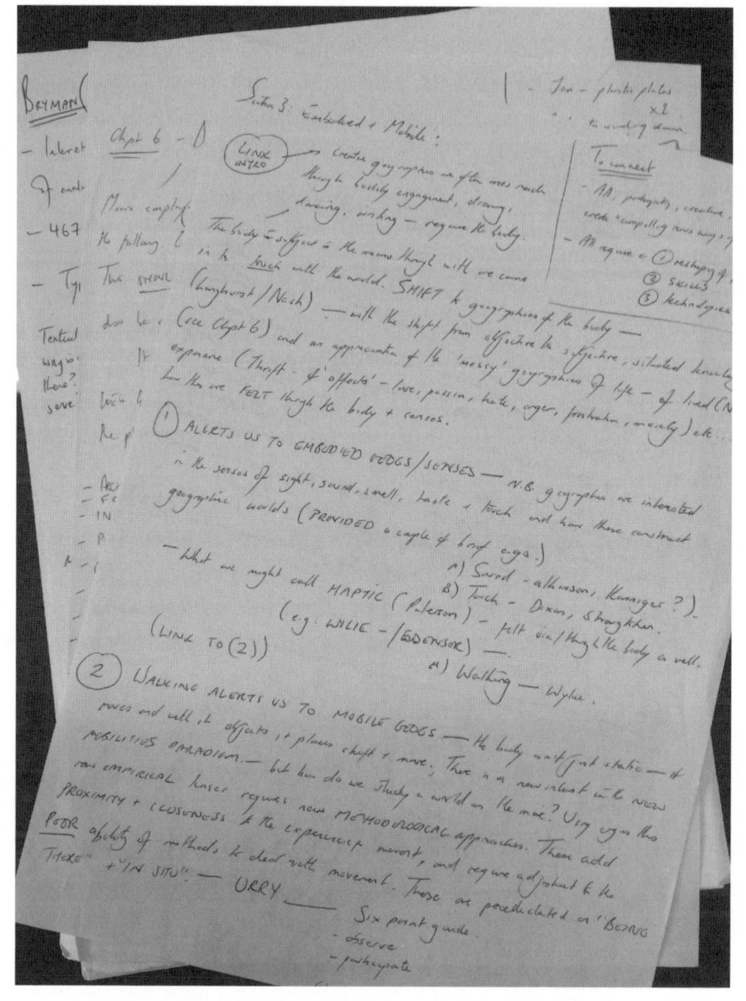

Figure 11.1 Planning ahead of writing: An example from the author. (Source: Kimberley Peters)

certainly be possible to structure the sections that makeup this text somewhat differently. Planning, then, is about settling on what you feel is the most logical structure for the piece. That said, as DeLyser rightly points out, writing is formative, and even though we might have meticulously planned the way our argument will unfold, it could change in the process of writing (2010: 427). We shouldn't be afraid to let this happen, and should rest assured we can return to the text later to see if revisions to our planned structure 'work' and still convey our intentions for the chapter effectively.

Finally, when we plan it is worth bearing in mind the word count of the dissertation. Often students are given an overall word count. Consequently, I am regularly asked by students 'how many words should each chapter be?' This is a difficult question to answer (and different supervisors might respond differently!).

That said, it is worth dividing up your words between the various chapters you have, but remember that the bulk of your words should be in your results, analysis and discussion chapter. This chapter is normally double the size of any other chapter (for a 10,000 word dissertation it will be almost half the word count, for example). The reason for this is because it is in this chapter/these chapters that you display your research and make sense of it. Try to use your rough word count as a guide when you plan and when you write. Sometimes we can be stuck for words. This is where careful planning can help us to populate headings with content. Sometimes, though, we have too many words (and it is surprising how often this happens, in spite of many student concerns about the length of a dissertation). Planning chapters often means deciding what to include, but also what to leave out. Given you have a limited number of words, some information will not 'make the cut' and planning helps you to think seriously about what is important, and what is more tangential.

When to write

It is one thing knowing what to write, but when should you actually start writing, and in what ways do deadlines and the need for drafting shape the writing process?

Write as soon as you can

Many would argue that you should start writing as soon as you can. It can be tempting to think that we should wait until we have finished all of our reading and all of our data analysis before we commit any words to paper. It is natural to have an anxiety that you might write something differently once your project is further along. It is also easy to make excuses: 'it'll be better once I've finished all my reading' – or 'I'll start when I've done my other assignments.' A dissertation happens in stages (as described in this very book). Therefore, writing can occur in stages too. If you put off writing, it can become a hurdle that might seem insurmountable later on. As the fiction writer Margaret Atwood notes, there is never a perfect moment to start writing:

> You always think, 'Oh, if only I had a little chalet in the mountains! How great that would be and I'd do all this writing...' Except, no, I wouldn't. I'd do the same amount of writing I do now and the rest of the time I'd go stir crazy. If you're waiting for the perfect moment you'll never write a thing because it will never arrive. (Atwood, cited in Hoby, 2013)

Try to jot down words as you go along. Writing early on in the process can be reassuring. Not only can it enable you to get into the habit of writing (which can help words to spill on to the page a little more easily), it can also be reassuring to get some words under your belt long before the final deadline. That said, when we write, it doesn't have to be chronological. Just because we write from the beginning of the process doesn't mean we have to start writing from the beginning of the dissertation. Few projects are written in order. DeLyser recommends that it is best to begin 'somewhere in the middle', often with a part of the dissertation you are 'confident or excited about' (2010: 426).

Write to deadlines

The question of when to write is also often determined by deadlines. A dissertation is not only a long piece of work in terms of the word count, it also takes place over a long period of time. As suggested here, writing from the 'get-go' is beneficial. Always be wary of leaving your writing to the last minute. Unlike other pieces of written work that you will have completed – which are often much shorter – a dissertation cannot be written overnight. In Chapter 1 a timetable was recommended as a tool to help you plan your time for completing the dissertation as a whole. It can be useful to write a further, more detailed timetable for the writing stage. When do you hope to have a literature review drafted? When do you plan to write up your main analysis and discussion chapters? In your final year you will often have to juggle writing up your project with other module work, so ensuring you use time between your other commitments effectively is crucial to success. But always be aware that whilst it is beneficial to plan your time, deadlines can be useful when writing (DeLyser, 2010: 426). Indeed, sometimes, 'with the minutes ticking away ... you'll find you *have* to write something. Whilst this is not the best strategy for creating truly well-crafted work ... a finished paper is better than no paper at all' (DeLyser, 2010: 426; emphasis in the original). It also worth noting here, that being behind and feeling behind are two different things. When we write, we often compare ourselves to others (Lamott, 1995: 123). You may see your friends steam ahead with their writing, whilst you feel yourself lag behind. However, remember that every dissertation is individual to the person completing it. Accordingly, students will often be writing different things, at different points in the process. Therefore, work to your own writing timetable and be confident in the structure it provides in enabling you to get the job done.

Write drafts and re-drafts

When you plan, it is also worth factoring in time to draft but also to re-draft the dissertation. Dissertations are very rarely written in one go. As DeLyser notes, if we 'hold ourselves to a notion that whatever we first write ... will remain that way, then we are indeed putting a great deal of pressure on those first words' (2010: 426). Some departments will have deadlines to read draft material. In other cases your supervisor may set a time they are available to look over draft work. Ensure that you know whether your draft material can be commented on, and what (if any) deadline there is. All earth-writers – from beginners to the most experienced of authors – have to draft and then re-work their writing. Few scholars (if any) will write the final version of something, at the first attempt. As the geographer Pauline Kneale notes, 'ask a tutor how often s/he rewrites before sending a piece to a publisher? "Lots" is the only answer worth believing' (2003: 180). To confirm this, it is probably worth knowing that this book has been written several times over before reaching its final form.

So how do we draft and re-draft? First we must plan and commit words to paper. Writer Anne Lamott calls this the 'down draft'. This is the first draft where we simply get 'something – anything – down on paper' (1995: 25). This first draft is crucial. It gets us writing and helps us start to formulate our thoughts. We have to be open to the fact that this draft might be far from perfect. Reassuringly, as Lamott notes, 'all good writing begins with terrible first efforts' (1995: 25). Once you have a 'down draft', open yourself to feedback. This can be daunting (DeLyser, 2010: 433). But as the author of one of my favourite books reflected, 'to begin with, I wouldn't let anyone read [it, but] I couldn't tell how good or how bad it was. It could have been excruciating! Gradually though, I let a few people read it' (Collins, 2005). Be confident in the knowledge that your supervisor will not expect a perfect dissertation at the draft stage. Feedback, however painful, is the means by which you can improve your argument. External readers (supervisors, peers or family members) will be able to pick out parts that might be unclear, or where you could expand your discussion further.

Once we have feedback, we should leave time for what Lamott calls the 'up draft' (1995: 25). This is the second draft, where you 'fix up' your writing, clarifying and bolstering your argument. For some students, this process can be frustrating as it means acknowledging the failures in our first draft. However, you shouldn't see feedback as negative but rather as a positive way in which you can strengthen your writing. Try to take any criticism constructively. As Limerick notes:

> A carpenter, let us say, makes a door for a cabinet. If the door does not hang straight, the carpenter does not say, 'I will not change that door; it is an expression of my individuality; who cares if the door will not close?' Instead the carpenter removes the door and works on it until it fits. That attitude, applied to writing, could be our salvation. If we thought more like carpenters ... we could simply work on successive drafts until what we have to say is clear. (Limerick, 2012)

In short, re-drafting always produces better writing. Indeed, the process doesn't end there. We must also factor into our writing timetable time for the 'dental draft' (Lamott, 1995: 25). This is the final draft were we 'check every tooth' (Lamott, 1995: 25) and ensure our writing is clear of errors, accurately referenced, and so on. It is important to leave enough time in the process of writing to complete a 'final' or 'polished' draft and we attend to this process in greater detail in Chapter 12 to follow.

How to write

So we know what to write and when to write, but how do we actually work out what we want to say (and how we want to say it)? Furthermore, how do we deal with the moments when we don't seem to know how to write?

Learning to write an argument

To work out what we want to say, we have to develop an argument (Information Box 11.2). As a written piece of work, the dissertation should tell the story of your research – beginning to end (DeLyser, 2010: 426). It should also, fundamentally, set out the research question you posed, and provide a response to that question.In order to develop an argument, you must allow yourself time to think (Kneale, 2003: 75). You need to look over your data (see Chapter 10), refer to your research question (Chapter 4) and consider how the two weave together in terms of what you will be able to claim. There is no easy way to do this but remember that you can be led by your data (what does it reveal and how can you use those revelations to make a case in relation to your research problem?), or you can be led by your ideas (what statement are you trying to make and how can you use your data to substantiate that statement?). Both are valid approaches, but both rely on evidence to make probable and robust conclusions (see Bonnett, 2014). Indeed, evidence is crucial to an argument and when you write you must back up what you say. Poor arguments are those that 'leap to conclusions' without substantiating that argument adequately.

INFORMATION BOX 11.2
MAKING AN ARGUMENT

What is an argument? The Oxford English Dictionary (2013) describes an argument as a process of reasoning that leads to a conclusion. It is a statement, or series of statements that should convince and persuade the audience as to their validity. To do so, an argument should be based on rigorous debate and substantial evidence. In your human geography dissertation you should construct an argument, based on careful consideration and strong use of data, that provides a credible response to the question you have posed.

When you consider your argument, be aware of the fact that there are different kinds of arguments you can make (DeLyser, 2010: 429). Some arguments will be bold, stressing a firm position in relation to a research question: the evidence shows, unequivocally, that residents on the east side of Sheffield are poorer than those on the west. Whilst it is vital to make an argument, it is also vital not to overstate your argument. Writing is the mode through which you convey your position and it is important to use words carefully. If your argument has limitations, do not exaggerate your claim. Other arguments will be more subtle: on balance, there is reason to believe that information technology can benefit development in rural regions, but this will depend on the social and infrastructural conditions of any given region. Some arguments will consider both sides of a debate (the 'yes, but' argument, see Kneale, 2003: 86). Here arguments will provide evidence that shows 'each side is valid' and 'each side has it merits' (DeLyser, 2010: 429). This kind of argument is not 'selling-out'. It is often appropriate when your project deals with complex geographical matters.

Finally, when you make an argument it is crucial that your argument *is* geographical and has 'relevance to geographers' (DeLyser, 2010: 429; see also Chapter 2). Human geographers often deal with topics that might also be addressed by other social science and humanities scholars. When interpreting evidence and developing an argument, some students can stray from the geographical significance of their work. When you form your argument (or indeed arguments, as a dissertation may seek to make more than one claim) be clear what contribution your debate adds for geographers. This often means referring not only to your data, but also connecting this data back to the geographical literature.

Learning to develop a style

Writing an argument is not only about committing claims to paper. It also matters *how* you make those claims. As noted at the start of this chapter, geographers are not explicitly trained to be professional writers. Yet write we must. The good news is that we can learn to write, and learn to develop a style that is suitable for the human geography dissertation being produced. As DeLyser has noted:

> We treat writing as if it were an innate talent, something we are simply able to do well – or not. Luckily that is not the case, for writing like carpentry, gymnastics and drawing, is only partially talent-determined. Like the other three, writing is also a skill and a craft. It can be learned and practiced, honed and sharpened, practiced some more and perhaps even nearly perfected. (DeLyser, 2003: 170)

We have seen already that drafting and re-drafting is a fundamental task when you write up your human geography dissertation. However, in this process you will also be honing a style that helps you convey your argument. Many students worry about the question of style but often, whilst writing, your voice will emerge. This voice will often reflect the tradition of your dissertation (as we saw earlier, some dissertations that come from a perspective of objective knowledge may be written in the third person, with chapters laid out differently from a qualitative dissertation that is forged from subjective knowledge and is written in the first person). The key is to ask what style is appropriate for the argument you are making. Returning to Task 11.1, consider the styles used in dissertations that use similar methods or ask questions akin to your own.

A dissertation will not lose marks if it is written in a conventional style, following a standard structure. What matters is that the style, structure and, importantly, content, convey your argument. That said, do not be afraid to adopt a creative style for your dissertation, if this suitable. Even for so-called conventional dissertations, creative techniques can be employed to engage the reader (remember that your work has an audience and you have the task of keeping them interested in what you have to say). Creativity might come in various guises. A few years ago, a student I was supervising came to me with an idea for writing up her dissertation. The student had focused their dissertation on the geographies of sport, and, in particular, the gender politics faced by female discus throwers. She was interested in the ways that such female athletes often failed to find a 'place' for themselves in popular media (sports television coverage to specialist magazines). The masculinised stereotypes associated with these throwing sports, and the often muscular physique of women, seemed to limit the reach of these athletes in the public domain (see also Johnston, 1996; Young, 1990). The student, committed to 'telling this story', decided to structure the dissertation like a discus throw. The introduction and literature review were titled 'the wind up' (the section where the athlete puts in the groundwork and prepares their throw). The methods chapter was titled, 'the spin' (the section where the athlete puts in the work that leads to the conclusion – the throw itself). The analysis and conclusion were grouped under a section heading called 'the release' (the section where the athlete realises the potential of the throw and finds out how long it is). For such a socio-cultural dissertation we can see here how this inventive approach both reflected the dissertation content and worked as an effective structure.

Whilst this was quite ambitious (and such 'plays' with structure are not always necessary or indeed desirable) there are other ways to be creative and grab the attention of the marker. DeLyser recommends, for example, using anecdotes in writing (2010: 428). Is there a short story, an interesting quote, a section of ethnographic data or a fascinating statistic that might serve to open your dissertation or chapters of your dissertation? You might also use 'metaphors and descriptive language to help evoke a feeling of a scene, a place or a person' (DeLyser, 2010: 427). You can often weave such creative and engaging techniques into your writing during an 'up draft' or 'polished draft' (see previous section).

Go online! Visit **https://study.sagepub.com/yourhumangeography** to access Sage journal articles that illustrate different styles of geographical writing. Ask yourself what kind of style works best in respect of your own dissertation. Also consider what creative and interesting techniques the authors have used to engage the reader.

Learning to write even when you 'can't'

All this talk of writing assumes that writing is an easy thing to do. But as other students (and staff for that matter) will tell you, there are occasions when writing just won't come. When this happens (and it does happen) it is useful to accept that you are not alone. Everyone experiences writer's block. As Anne Lamott notes, 'writer's block is going to happen to you' (1995: 176). Sadly, she is right. There have been days with this book when the words have formed readily, and days where a blank page has stared painfully back at me. But how can we train ourselves to write in all circumstances (whether we are feeling the potential or not)?

The author Philip Pullman expresses a common approach to dealing with the writing when things get hard. He states:

> The most valuable thing I've learned about writing is to keep going, even when it's not coming easily. You sometimes hear people talk about something called 'writer's block.' Did you ever hear a plumber talk about plumber's block? Do doctors get doctor's block? Of course they don't. They work even when they don't want to. There are times when writing is very hard, too, when you can't think what to put next, and when staring at the empty page is miserable toil. Tough. Your job is to sit there and make things up, so do it. (Pullman, n.d)

This approach is a useful one for many writers. If we remember that the job of the geographer is to write about the earth, we can appreciate that this is a task we have to do. Often it can help to write regularly (Murray, 2007). Writing something daily (as short as a paragraph or as long as a chapter section) can help you to get into a routine. If you fail to find a routine, several techniques can also be helpful for easing the path back to writing (see Information Box 11.3 and Task 11.2). You can also find advice from articles that can be accessed via the Companion Website.

Go online! Visit **https://study.sagepub.com/yourhumangeography** to find links to two useful articles from the Guardian Higher Education pages which consider how to deal with those moments when the writing just won't come. These articles complement Task 11.2.

INFORMATION BOX 11.3
MIND MAPPING

Rather than trying to construct prose, instead make jotted notes of what you want to say. This can provide 'thinking space' that can then help you work out what it is you are trying to write. We often map out an entire chapter, but when words fail us, it can be helpful to map out individual sections or even paragraphs of a chapter.

(Continued)

(Continued)

Write down all you know about X

When we don't know what to write, or how to start, it can be useful to just write anything down on the given topic. Say, for example, you need to write a section of your dissertation on environmental policy, it might help just to write down all you know, and then use that 'down draft' to craft it into something more logical.

Move on to something else

Writer's block can sometimes occur as we struggle over a particularly tricky part of the dissertation. The stubborn part of most earth-writers will want to try and carry on regardless. However, it can be more beneficial to move on to another section of the argument – perhaps one that is easier to construct. If we are able to write this smoothly it can help us get into a flow for returning to more difficult sections.

Freewriting

This is a process of writing something, anything, as a means of getting started. If your writer's block is particularly debilitating, this is a good place to begin. Freewriting is concerned with putting words on a page. For example, you might open a word document and simply write about your weekend. In committing these words to a blank page you have begun writing, and this in turn can make the task of writing 'on topic' easier to return to. (Taken from Murray, 2007: 173; based on Boice, 1994.)

TASK 11.2

When faced with writer's block, jot down answers to the following questions, which are targeted at helping you write up your human geography dissertation. Try to write the longest answers you can, and add as much detail as possible. Don't worry about whether the sentences or paragraphs are fluent and grammatically correct. There is no need to add references or consider style. At this stage, just write!

- What was the inspiration for your project?
- What piece of reading grabbed your attention most on this topic? Why was this?
- What has been the most interesting thing you have found out? Why?
- What surprised you the most? Why?
- How would you describe your project to family and friends?

All of this said, sometimes the stress of completing a dissertation can be overwhelming and hints and tips for writing can seem a hurdle too. If this happens, speak with your dissertation supervisor or use resources available at your institution. Your Student Union may have drop-in sessions at key times of year offering advice for dealing with workloads and deadlines and almost all universities will offer student counselling services.

The 'where' of writing

As geographers, we invest a lot of time in thinking about space and place. Indeed, we are often very attuned to the difference that place makes to our sense of well-being or unease, comfort or discomfort, enjoyment or fear. Accordingly, it makes sense that knowing where you work best can help in writing up your human geography dissertation. It goes without saying that you should try to write somewhere that is productive for you. Some students work best when surrounded by others. In the past I have known students to organise writing groups, where a cohort will get together to write their independent projects. The group dynamic can be helpful as each individual is spurred on by the others in the group to get the work done. Other students will write best on their own.

However you choose to work, be it on your own or with others, consider what will be the most productive environment. Typically it is best to select a place free from the things that distract you (and what is distracting will be different for each person). It might be that the 'best' place is the library, another university study space, your bedroom, or even a coffee shop. Be aware of how you work best and capitalise on those conditions. If you know you need silence, regularly take yourself to a quiet study space on your campus. If you like to work to music, consider where you can do this, so you can have it in the background. That said, DeLyser (2010) urges us to be cautious of investing superstition in our preferred places and times to work (a particular seat in the library or a set time in the evening). These rituals can provide comfort but good writing can just as easily come from a different library desk or another time of day. Likewise, it can often be helpful to change locations (especially if you have writer's block). When trying to finish this book, I took the rather drastic measure of working at a university in New Zealand, carving out dedicated time and space to finish the manuscript. So if you are staring at a blank screen, take a walk, swap seats, try working somewhere else. This can often help as it provides a break to the conditions that might be limiting your writing (Walliman, 2011: 321).

Finally, you must consider where to save your work as you write it (Walliman, 2011: 321). It seems like an obvious point, but it is crucial to save your work and to do so in more than one place. For starters, make sure you save your work at regular intervals. You can often set your computer to do this for you. That way, should your computer crash, you will have a fairly recent copy of what you have been working on. Secondly, make duplicate files. Save your work on your computer, but also on a flash drive/memory stick. There are also many online modes of saving work too – including cloud storage or saving your work on email. Whatever you do, know where your work is saved and ensure you keep multiple copies of files in case things go wrong. Where you have written plans or notes on paper, make sure you keep these somewhere safe and, if you are particularly worried, take a photo or photocopy of plans so you have them backed up should they go astray.

Chapter Summary

- Geographers are 'earth-writers'. Writing is therefore a crucial means of conveying what we have found about the world through our research. A dissertation is typically composed of several chapters (introduction, literature review, methods, results, analysis and discussion, and conclusions) but its exact structure will depend on the tradition of your work. It is vital to plan each chapter carefully to guide your writing.
- Writing is a major task and it is useful to start as soon as you can so as to get into a habit of translating thoughts to paper. Writing should also reflect the deadlines you have (for a final copy and for any drafts). Furthermore, writing should be staged, with phases for drafting and re-drafting in order to polish your text.
- The quality of your writing depends on the content and your ability to make an argument that is substantiated with evidence. A good-quality argument is also one that is stylistically well-written and engaging. This is often easier said than done, so if you are faced with writer's block, use various techniques (see Task 11.2) to inspire writing.
- It matters where we write (as well as what we write and how we write it). Write somewhere you are comfortable and that best suits the way you work. This might be in a group, or somewhere quiet on your own. Make sure you save your work and back it up in multiple places.

Key readings

Bonnett, A. (2014) *How to Argue*. London: Pearson Education.

DeLyser, D. (2010) 'Writing it up', in B. Gomez and J.P. Jones III (eds) *Research Methods in Geography: A Critical Introduction*. Oxford: Blackwell. pp. 424–36.

Lamott, A. (1995) *Bird by Bird: Some Instructions on Writing and Life*. New York: Anchor Books. (This easy-going text is a reassuring and entertaining guide to the highs and lows of writing.)

Walliman, N. (2011) *Your Research Project: Designing and Planning Your Work* (third edition). London: Sage (see particularly Chapter 7, 'How to Get Started with Writing').

12

THE LAST HURDLE: FINAL CONSIDERATIONS

CHAPTER MAP

- Getting started with finishing
- Inserting the bookends
- Proofing, polishing, printing
- Marking and results
- *Being* a geographer

Getting started with finishing

For Tim Ingold, life is lived along a line (2007). In his extensive writings concerning space, place and time he has contemplated, again and again, the ways in which we experience the world (see also Ingold 2011a; 2011b). It is by journeying along paths, or lines, he argues, that things *happen*. Your human geography dissertation can be thought of as a line, or a path that you have followed as well as constructed. Indeed (as you can probably confirm), the dissertation is a journey. Along the way you will have invariably travelled to different places both physically (the library, your supervisor's office, various research sites) and emotionally (to moments of enjoyment, surprise, despair or frustration). There will have been times along the path when the research experience felt exciting and rewarding. There will have been other times it felt overwhelming or perhaps even dull. The dissertation process is rarely a straight line. Clearly it has ups and downs.

Although this book has followed an obviously linear form, it has also ventured that the dissertation is a piece of work that shuttles back and forth, with highs and lows. Certainly a dissertation is designed, 'done' and delivered, but each part invariably maps on to another, linking backwards and forwards to other stages in the process. Questions re-emerge in analysis. Methods map on to design. The line your dissertation has followed will have been complex. As you have likely experienced, paths are rarely smooth and seldom lead where they are meant to. As Ingold notes, lines can become messy. Knots can form. We may have to work through these, unpicking them before we can proceed. Threads can also fray, leading lines to splinter apart in all kinds of directions. Likewise, on your path to completing a dissertation, sometimes the journey will have become muddled. Sometimes it will have veered off track. But nonetheless, it is along lines and paths that Ingold notes 'people grow into a knowledge of the world around them, and describe this world in the stories they tell' (2007: 2).

In this chapter we consider how you conclude your own story, overcoming the last hurdle to reach the finish *line* of your human geography dissertation. We begin the end of this journey by considering the additional pages that 'bookend' your project. These are the preliminary pages that frame the beginning of the document and the concluding pages that round off the work. Following this, we track through the process of finalising your document for submission, exploring the ways in which you can best produce a professional-looking dissertation. In the penultimate section, the mystery of the marking process is discussed in order to shed light on what happens once you've submitted your dissertation. This section also attends to how to deal with results, be they good or bad. Finally, the chapter considers what happens next and the usefulness and relevance of your dissertation once it is 'delivered'. Indeed, a completed dissertation often continues to resonate long after it is finished. To return to our earlier analogy, the dissertation is like a line. It is 'open-ended' and leaves 'trailing ends' (Ingold, 2007: 169). It can lead to other paths – to future research, new interests or to employment opportunities. So for now, let's continue.

Inserting the bookends

In Chapter 11 we saw that dissertation projects typically consist of a set of core chapters: an introduction, literature review, methodology, results, analysis and discussion, and conclusion. However, this main block of writing is often bookended with a collection of opening and closing pages that are essential to your human geography dissertation. These bookends frame the project – providing both context and finality to a professional-looking dissertation. In what follows, we attend to each of these crucial components, beginning with the preliminary pages, and ending with the concluding ones. These are outlined in the order they typically appear in student projects. That said, particularly in respect of opening pages, the sequence of these pages may vary depending on your own institution. Following Tasks 1.1 and 11.1, check your own institutional guidelines – and previously submitted dissertations – to ensure you have these arranged correctly.

Preliminary pages

Title page

It goes without saying that your human geography dissertation should have a title. Often you will have given your study a title at the design phase of planning the project. When you reach the final stages of your dissertation it might be the case that your title requires minor amendments. This could be because the project changed slightly from the proposal or – with greater knowledge of the topic and your findings – you will want to tweak it to best reflect your study. Either way, your title should indicate your topic area clearly. Good titles are typically concise, (the point of the study can be lost in a long title); well defined (it should precisely reflect your study); and engaging (it will be the first thing the marker reads). As Hay usefully notes, a good title is 'functional and informative' – for example: "'Social consequences of homelessness for men in Adelaide South Australia (2000–2012)'". A bad title on the same topic could simply be "'Men and Homelessness'" (Hay, 2012: 35). Titles can sometimes be difficult to get right so don't be afraid to draft and re-draft your title until you are happy that it reflects your project well. Don't forget to add your name, student number, degree scheme and/or module code (as appropriate) to your title page.

USEFUL TIP

Your human geography dissertation will contain more than just an overarching project title. Individual chapters will have titles (and you may wish to have more inventive chapter headings than the obvious ones usually employed), and certainly sub-sections of chapters will have subtitles. Just like your main title, any title that appears in your dissertation should be clear and provide a useful description of what the chapter or sub-section contains.

Declaration

Most departments and schools will ask you to insert a page into the start of your human geography dissertation that states that the study is your own work and that any work which is not your own has been duly acknowledged and cited (for example, where you have referred to academic literature, that this has been appropriately referenced). Some institutions will provide a declaration as a template that you simply print off, sign and insert. Others will require you to word your own declaration and insert this. Make sure you check the rules of submission where you study. Also – take the declaration seriously. Universities will be unforgiving if your study has engaged with what is often called unfair practice (for example, plagiarism). Ensure you can declare that what you submit is *your* human geography dissertation.

Abstract

All dissertations will feature a short abstract at the start of the project (of 150–250 words, though, again, check your own institutional guidance). When I am asked by students what an abstract is, and how to write one, I typically describe them as a spoiler. We have often heard sports commentators or entertainment correspondents ask us to turn away from our TVs if we do not want to see the results of sporting events or showbiz competitions. If your family or friends do not want to know what your dissertation is about until they read it in its entirety, they should equally look away from your abstract. An abstract for your project should concisely describe your project. In a single paragraph you need to condense its focus, the methods you used, the results you came to and your conclusions. Task 12.1 helps you develop the skills needed for writing an abstract.

TASK 12.1

If you are in any doubt as to how to write an abstract you won't have to look too far for guidance and inspiration. Academic research that is published in journals will feature an abstract at the start of the article. Pick a journal that is relevant to your own project (see Information Box 3.1, pp 46–7, for a reminder of key geography journals). Review 3 or 4 abstracts from papers in the journal. Ask yourself how the author has constructed the abstract to provide a sense of the article. Did they use a clear opening sentence that summarised the focus of the study? Was there a linear structure that walked you through the key points? Were the findings summarised and a conclusion or contribution stated?

Now try to write your own abstract. Once drafted, ask yourself if it reflects the work you completed. If not, re-draft it. Once you are (more or less) happy, ask a peer, friend or family member to read it, and then explain to you what your project is about. If they are able to tell you the key points of your study, you know you have succeeded in writing a clear abstract. If not, you may have to tweak the abstract, or overhaul it entirely, before repeating the process.

Contents pages

Flick to the opening pages of this book and take a look at the contents pages. They are likely pages you have visited frequently when navigating the text and finding the sections or chapters that are relevant to you at any given time. That is the job that good contents pages serve. They should provide an overview of what is contained within a text and signal where the reader will find particular sections of the dissertation with corresponding page numbers. Indeed, make sure that the text of your dissertation has continuous page numbers from start to finish. It is typical to start numbering your pages from the introduction chapter, with preliminary pages (such as the abstract and contents pages) indicated with roman numerals. Again, the opening pages of this book provide an example. It is up to you how best to organise your contents page. A basic contents page will simply contain the chapter headings and page numbers. However, it can be beneficial to include the sub-headings used within chapters to provide more information for the reader. That said, it is also important to make sure your contents pages are accurate, especially if you are compiling this by hand (some word processing packages can compile a table of contents for you using the 'headings' styles). When drafting (and re-drafting) on a computer, page numbers can change frequently, so complete your contents pages at the end, when your document is finalised.

USEFUL TIP

In addition to a table of contents that displays information about the written text, you may also need to include pages that indicate where figures, tables or graphs may be found. If your dissertation uses images or maps (typically called 'figures'), provide a list of these with corresponding page numbers on a separate page. The same goes for tables and graphs. Hay also notes that if your dissertation uses lots of abbreviations or acronyms it can be useful to include a preliminary page that lists these for the reader to refer to (2012: 40).

Acknowledgements

Students often tell me that the acknowledgements are the part of the dissertation they enjoy writing the most, mainly because these are typically written with the acknowledgement that you have finally finished your project. It is usual to write a brief 'thank you' when you reach the end of your dissertation to those who have assisted in its completion. Although not mandatory, many student projects include a short paragraph thanking those who have made the project possible. Whilst this often includes those who have been a part of your immediate support networks (family, friends and supervisors), acknowledgements are particularly important in respect of those who gave their time to help in the research process (departmental staff who helped you draw maps; research participants who gave up their time to be interviewed; archivists or librarians who assisted you in finding textual records, and so on).

Concluding pages

Appendices

The use of an appendix (singular) or appendices (plural, where you might have appendix 'A', 'B', 'C', and so on) is often a grey area for students. When do you include them and what do you include? Firstly, an appendix is not a formal requirement. Not all dissertations have or even need appendices. Where they are useful is if you want to include supplementary material that supports your dissertation. For example, you may wish to include interview transcripts or an example of a survey you completed to show the marker the work you have done. It is worth remembering that appendices do not have to be read by the marker (although most will browse through these). Do not hide your best data or crucial conclusions in an appendix. It should also not be a 'dumping ground' for material that won't fit elsewhere. Rather, it should be another well-crafted component of your project that provides additional evidence (Hay, 2012: 47).

Reference List

A reference list can be thought of as a different kind of acknowledgement – one that gives credit and recognition to the work that has supported and shaped your own human geography dissertation. As noted in Chapter 4, reading is fundamental to your dissertation. Each study will be informed by the geographical work that precedes it. A reference list appears at the end of the dissertation, providing a list of the sources used in the text (a 'bibliography', by contrast, contains sources that have influenced your work, but which are not cited in the text). Accordingly, the reference list should contain all of the sources cited in your dissertation. If you refer to a point or idea made by someone else and choose to use their words (quoting them directly) then clearly you must cite and reference them. Likewise, if you refer to a point or idea made by someone else, but do so in your own words (paraphrasing) you must still cite and reference the author. Any source of information that is not your own must be duly recognised. If you do not reference other people's ideas or words, you are effectively (whether you intend to or not) passing them off as your own ideas. This is plagiarism. It goes without saying that such unfair practice is unacceptable in all institutions. Almost all schools and departments will ask you to submit your dissertation electronically, as well as in paper form, and it will likely be checked with anti-plagiarism software. Do not run any risks, or, with the stress of a dissertation, be tempted to pass someone else's work off as your own.

There are differing ways to reference but the Harvard System is most common. Here the *author*, *date* (and *page number*, where appropriate for direct quotes) is cited in the main text when the work of someone else is referred to. A full reference for the citation is then found in the reference list, where it will be listed with other sources used, in alphabetical order. Typically, by this stage of your degree, this will be a common and straightforward practice that you engage with automatically. If you are still unsure of referencing practice, or how to refer to specific sources, links to good reference guides can be found on the Companion Website. However, it is always best to consult your own department's style guide to ensure accurate referencing. Also see Information Box 12.1 below for tips on citing and referencing professionally.

Go online! Visit **https://study.sagepub.com/yourhumangeography** for web links to comprehensive guides to referencing. These are specific to the institutions they relate to, but they provide good general information. Be sure to refer to your own institutional guidelines (see Task 1.1 in Chapter 1).

INFORMATION BOX 12.1
GETTING REFERENCING RIGHT

The following tips help ensure you have a tidy and professional reference list:

Use only one system for referencing

As noted there are different ways to reference. Although Harvard is common, other styles are used (historical geographers will often use the footnote /'Oxford' system, for example). Check with your department, then pick one system and stick with it. Do not muddle up two different ways of referencing.

Use a consistent format for citations and references

If you look at books and journal articles you will notice that the citations and references all look the same. You should format your citations and references consistently too. In other words, for in-text citations, if you place a comma after the author name – for example (Clark, 2015) – make sure that comma is on *every* citation to follow. Likewise, in the reference list, if you use full stops, italics, underlines, or bold text, make sure you do so on each reference. For example, if we were to cite a journal article like this – Author, Initial (Year) 'Title of Article' *Journal Title* **Volume Number**: page number – then every journal article to follow should employ the *same* style. I mention this because it is now easy to cut and paste references from the internet (especially using Google cite). These are rarely formatted to the same style and if inserted straight into your document, you will find you end up with a sloppy, unprofessional reference list.

Reference online sources accurately

Producing a professional reference list means accurately formatting the different kinds of sources you might use. Whilst books and journal articles can be quite straightforward, online sources can often be a cause for confusion. Essentially, if you use a webpage (for example, a news story) your reference should include the full URL, or web address, for the source. The date you accessed the source should also be added. If there is an author, title and date for the piece, use these details to construct the reference as you would any other. If these details are not evident, you can organise the reference by the title of the website, the name of the page/article and, if no date is present, use the term 'n.d.' (meaning 'no date'). In general, third-party material hosted on personal websites (for example scans of articles or book chapters) should be avoided. In this instance, refer to the original source.

Ensure your references are complete

As mentioned, it is now easy to cut and paste references from the internet (rather than having to type them out). Sometimes, however, these references will be incomplete. I am regularly circling incomplete references on student projects and essays which I mark, where a journal title is abbreviated, or the place of publication is missing from a book reference. Again, this leads to a poorly constructed reference list. If you do cut and paste, go back to the reference and ensure it is complete. Don't trust that it will be. If in doubt, use the old-fashioned approach of writing the reference out for yourself.

(Continued)

(Continued)

Only put sources in the reference list which appear in the main text

When marked, staff will check your referencing for accuracy. When I do this, it tends to be because a student has cited something that I'd quite like to read and I am keen to find the full details in the reference list. It is frustrating when a reference is not in the list. On the other hand, students can sometimes try to make it appear that they have read more by inserting extra references into the reference list. This is not a wise move. It will be noticed. Accordingly, make sure your in-text citations and reference list match.

Use page numbers when appropriate

It is worth remembering that for in-text citations where you have used a direct quote, you must refer to the page number of the text you have used (if one is available). The only time when this is not feasible is if you are using a source that has no page numbers, such as a website. In these cases, students can indicate that no number was available by writing 'n.p.' (short for 'no page'). It is also worth remembering when page numbers are required in the reference list. Journals are a good example, but a full page range is also needed where chapters from a multi-author volume have been cited. The omission of page numbers for such sources is surprisingly common.

Proofing, polishing, printing

Once you have your main chapters in their final form and once your bookends are in place, you need to think about the finishing touches that complete the dissertation. A completed dissertation is one that is proofed and polished – that is checked for errors and honed as much as you can within the time available – and one that is printed and bound to the specifications set by your school or department. In Chapter 1, I asked some of my own students what they wished they had known at the start of the dissertation process. Now, in this last chapter I turn to their advice again to find out what they wished they had known at the end of the journey. Here's what they said.

'I'd say to read it and then read it again – I kept finding spelling mistakes!'

Whilst we have finite time available to finish our dissertation, it is worth allocating some time to proof read your work (and, as my student recommends, to then proof it again). It will almost certainly be the case that there will be errors with grammar, spelling and/or punctuation. And – as almost all books that attend to academic writing will tell you (see Hay, 2012; Osmond, 2016; Walliman, 2014) – it is vital to get these components right. A dissertation that has weaknesses in writing will struggle to convey accurately what you want to say.

DeLyser recommends reading your dissertation aloud to yourself (2010: 432). You are more likely to pick up any errors in this way. I would also recommend not relying too heavily on the spell-check tool on your computer. Be cautious in accepting recommended changes (especially to names which it may then spell incorrectly). In addition, spell-check will not pick up on errors to writing that are not misspelt (for example an incorrect use of 'there', 'their' and 'they're'). Accordingly, run this as a first review of your text, but then scrutinise your work more thoroughly yourself or – better still – get a family member or friend to read the text carefully for you. Also be aware that your institution will likely offer student support services that you may draw upon too.

Key reading

Osmond, A. (2016) *Academic Writing and Grammar for Students* (second edition). London: Sage.

'Get your presentation right. I was really pleased with how mine looked in the end'

With so much time spent on writing the dissertation, the presentation of the report can often take a back seat. Yet it is worth spending a little time considering how to present your work. Whilst it is the content of what you have written that will gain you the most marks, usually a small percentage of your grade will be determined by presentation. A well-presented dissertation not only says something about your approach to the dissertation, and the time you have dedicated to it; a polished appearance can also help communicate your project findings more clearly.

For starters, check your own school or departmental guidelines in case there are particular specifications for the presentation of your work. For example, some institutions require projects to be in a minimum 12-point font, double spaced and printed one-sided. Other institutions may have different rules. Make sure you are aware of any presentation guidance and follow this closely.

Second up, consider the layout of your text. You will often have more freedom here so work out how to best arrange your project so it looks professional. For example, it is good practice to start all new chapters on a fresh page. It may also be a neat touch to add running heads that indicate the chapter number and title. For most students, this will be the one piece of work that you have dedicated most time towards. Getting the 'look' of it right ensures you will have produced something you are proud of.

Key reading

Hay, I. (2012) *Communicating in Geography and the Environmental Sciences* (fourth edition). Oxford: Oxford University Press (see particularly Chapter 2 'Writing a report').

'Make sure your images are of good quality and you can see all the captions'

As geographers, we tend to use a lot of visual sources in our work. Accordingly, if we have figures, tables, maps, graphs or sketches in our own dissertation we must ensure they are of good quality and that when we print our work they are a decent size and clearly formatted. For images such as photographs or sketches, it is generally good practice to insert these as files rather than cutting and pasting them into a document. This keeps them in their original file state. For photos in particular, crop and edit them before insertion, rather than stretching them once in a Word document (as this will distort the image). For sketches or field diary materials that you want to display, scanning the image is often the most appropriate way of then integrating it into your dissertation.

For graphs, tables and maps make sure that your choice of graph, table and map is appropriate. A badly selected graph – for example choosing a bar graph when a scatterplot is more suitable – can make 'the difference between results being meaningful or not' (Field, 2010: 323). Moreover, make sure that the written information on graphs, tables and maps is legible and clear, that labels are present and colours allow the data to speak (see also Chapter 10). If you cannot include colour prints, check that your colours can be clearly distinguished in greyscale. It is good practice (albeit a little more costly) to print images before the final submission to check their quality in advance.

Finally, you should consider how visual sources fit into the dissertation so it flows well. Inserting graphs, images, tables, and so on, will often shift the writing in your Word document and you could be left with gaps and blank pages. You may wish to display images on separate pages (in landscape or portrait orientation as appropriate). Alternatively you may wish to wrap text around images so there is not an empty space after a paragraph and before an image appears.

Key reading

Field, R. (2010) 'Data handling and representation', in N. Clifford, S. French and G. Valentine (eds) *Key Methods in Geography* (second edition). London: Sage. pp. 317–49.

'To be honest, I should have left more time for printing and binding – I was so stressed with all the queues in the library!'

A dissertation doesn't end when the document is complete on your computer or laptop. Although you will likely have to submit an electronic version (converting the file to a PDF is often the best way to do this to reduce the file size for uploading), most institutions still require you to submit one or more paper copies. Again, check your school or departmental guidelines on whether you have to submit one or two copies. You may also want to have a copy yourself (although often if you submit more than one, a copy will be returned to you after marking).

Also check the specifications for binding. Does your institution ask you to use a particular coversheet? Is there a preferred type of binding (comb binding or strip binding)? Also make sure that your left-hand margin is an appropriate width. Binding will reduce the space along the left-hand side of the paper, so if you do not adjust your text it will not appear in the middle of the paper when the document is bound and complete.

Also, as my student recommends, leave time for printing and binding. As Walliman notes, 'pressure on printing and binding facilities will become greater as the … [deadline] approaches (2014: 246). You will have to print and bind your dissertation in advance of the deadline, so, to avoid stress, work to complete your project a few days before the final date so that you can avoid the queues (or get up early to get yourself in the queue first!).

Marking and results

Submitting your work can (and arguably should) be a time of celebration. Indeed, it is often worth observing the final submission with some kind of reward for your hard work. Whether you have been happy or not with your progress and the work you handed in, completing your dissertation is often a significant personal milestone as well as bringing you one stage closer to the end of your degree. That said, for some students it can feel odd to no longer work on the dissertation, which invariably becomes something of a constant in the lives of many third- and fourth-year students. Many students prefer to just plough on with further assignment work or exam preparation. Whatever you do, you will have achieved a great deal in submitting your work and it is worth remembering that fact. However, few students really know what happens next, between the stage of submission and the release of marks. In this penultimate section I want to demystify the marking process and explore how you deal with your results when they arrive.

Understanding how it is marked

The criteria

The best way to understand how your work will be marked is to look at the criteria for the dissertation project. As noted in Chapter 1, it is useful to consult this right at the beginning of the dissertation process. You can only ensure your project ticks the relevant boxes if you know what those boxes are. The exact nature of any dissertation marking criteria will differ depending on your institution. That said, most schools and departments will mark you on the standard of a similar set of components when reading your dissertation. These are outlined in Information Box 12.2 below. When staff review your work they will closely follow the marking criteria to make their evaluation. Depending on how well you meet the criteria, the marker will place you in the brackets common to degree level work (in the UK these are the first category (70 and above); the second, 2:1 (60–69), and 2:2 (50–59); the third category (40–49)). Indeed, the criteria will normally be split into these categories, specifying what level of engagement is needed for each. Accordingly, the criteria are essential in understanding how a mark is reached.

INFORMATION BOX 12.2
MARKING CRITERIA

You should always consult the marking criteria specific to your own institution. That said, it will likely contain wording that covers the following bases, against which the dissertation is evaluated:

- **Research question and rationale** – does your dissertation have a clear research question and set of aims and objectives? Is there a justification for the research?
- **Research context** – is your dissertation adequately situated within the ideas from the academic literature that have inspired it?
- **Appropriateness of methods** – in terms of the question you have asked, are your methods suitable for collecting data relevant to answering that question?
- **Use of methods** – have you provided a thorough and critical discussion of how you employed your methods and discussed the limitations of your research?
- **Quality of analysis** – have you used an appropriate approach and suitable tools to make sense of your data? Does it help you to interpret your materials?
- **Credibility of findings and conclusions** – have you built a convincing argument based on your analysis and have you reached a concluding statement on what you discovered?
- **Literacy and presentation** – is your dissertation free of writing errors, well structured, clearly signposted and neatly laid out?
- **Standard of referencing** – have you cited the sources you have used, in full, in a reference list? Is this accurate and consistently formatted?

The marking process

Although all academic work is carefully marked, it is worth knowing that the dissertation is perhaps the most rigorously marked assignment you will write. This reflects the effort you put into the project and the bearing that the dissertation has on your final degree grade. It is typically a heavily weighted piece of work. Accordingly the dissertation is normally double-marked. Typically markers will read the dissertation carefully, against the marking criteria. Once each marker has arrived at a score they will meet to discuss the work. Here they will compare grades and agree on a final mark based on the evaluation of both parties. In my own experience, markers tend to make very similar assessments of dissertations (which should be a comfort for students to know!). On rare occasions where markers disagree and there is a discrepancy, a third marker is normally introduced to assess the work further. This process explains why the marking of your dissertation will seem to take a long time (typically much longer than you'd wait to receive marks for a standard essay). Whilst waiting can be difficult, it is necessary so that staff can give due time and attention to evaluating your dissertation.

Dealing with the results: The good, the bad, and the 'average'

During the course of this book I have taken (what is I hope) a very positive approach to completing your human geography dissertation. But it is also worth acknowledging that things do not always go the way we want them to (see also Parsons and Knight, 2005). I have to confess that I have been least looking forward to writing this section of the book. Writing a dissertation invariably comes with highs and lows. In Chapter 1, it was recommended that you should meet with your supervisor regularly. If things at any stage go awry – you cannot settle on a feasible research question; you can't access your datasets; your participants won't respond, so you have no data; you can't seem to write up your results – discuss this with your supervisor. Although it is your dissertation, supervisors have lots of experience from their own research work and from supervising previous dissertation students. They will be able to guide you as best they can.

That said, whilst it is possible to salvage a situation (lack of data, answers that don't respond to the question, and so on), once the dissertation is submitted there is little left that you can do except await your mark. The highs and lows that occur throughout the dissertation process are also a feature of dealing with the results of such work. Students often fall into several groups when marks are released. For some (and I hope many), they will be pleased with the mark gained. For others it will be a relief – the mark will be higher than they hoped. For others there will, unfortunately, be a sense of disappointment. It is surprisingly common for students who get good marks to be disappointed. Some want or hope for higher. Others feel their mark is 'average' when, in fact, it is very good. At times like these it is worth remembering how well you have done to merit the mark you have received (even if it is lower than you wished for). If this is the case for you, I can only urge you to remember what a tough job it is to finish a dissertation. Becoming a geographer is not easy (and it isn't meant to be either). The dissertation is a test of a range of academic and professional skills and these are duly challenging in order to develop your abilities (see Chapter 1). Finally, for one batch of students there will be a very strong sense of disappointment and frustration if the marks received are especially low. Sadly, marks towards the lower end of the spectrum do happen – and they can be both expected and unexpected. The only advice I can provide, should such an occasion arise for you, is to give the news time to settle, and to understand why that mark has been awarded. This means gaining feedback.

Gaining feedback

Often we are most concerned with the numerical mark that our dissertation has been allocated. But it is vital to understand how that mark was reached – whether you are pleased with the result or not. Feedback can help you understand both the strengths and weaknesses of your project (indeed, even the best projects usually have room for improvement). I would always recommend gaining any feedback available, be it in written form or via a meeting with your supervisor. Every school or department will offer dissertation feedback (albeit in different guises). Check your own institutional guidance for how you can gain feedback for your human geography dissertation. This will provide an explanation of why you received your mark. This will be especially helpful if you were disappointed with your result. It isn't always easy to accept criticism about our work but (at the risk of sounding condescending) it is crucial in the process of accepting why things turned out the way they did.

In Chapter 3 I noted that there is something that binds academic staff and students together. All academic staff (your lecturers, personal tutor, dissertation supervisor, and so on) will have once been in your shoes, designing, doing and delivering a piece of geographical research for the first time. They (we) also have a lot of experience in dealing with criticism and disappointment when writing up our projects. Professional geographers have to face harsh evaluations of their work regularly. When we submit our research for publication in journals or in book form, it is stringently peer-reviewed. As most staff will attest, some feedback can be particularly upsetting, especially when we have put our heart and soul into an article or book chapter. At first we might not agree with the assessment of our work. We might be annoyed that the reviewer has not understood, or that they have failed to see the toil we have put into our study. These feelings are often shared by students if they gain a disappointing dissertation mark. But as academic members of your school or department will also tell you, feedback, upon reflection, is almost always helpful. Whilst feedback may seem unhelpful at the end of your degree when you cannot take any advice forwards to future assignments, you can take it forwards as you move beyond university. With feedback we can become better geographers, job candidates, employees and students because we are open to hearing the opinion of others and taking such advice on board.

Being a geographer

As we reach the end of this book it is useful also to return to the beginning. In Chapter 1, I noted that completing your human geography dissertation was a part of a journey towards *becoming* a geographer. Geographers, we noted, are 'earth-writers'. They help author knowledge that assists in understanding the world. As you come towards the end of your human geography dissertation it is worth asking what you can do next with the work you have completed. The dissertation – and the work you have put in – does not only amount to the document you have produced. The dividends of a dissertation will be both personal (certainly you will have gained skills and understanding during the course of your study) but they may apply more broadly to the people you worked with and/or the places you have researched. Will Wright describes the benefits that have come from completing research, in the Graduate Guidance box to follow.

FUTURE TRAJECTORIES AND LIFE AFTER RESEARCH
WILL WRIGHT

My undergraduate dissertation represented a real turning point in my interest in geography. Based at the University of Sheffield, I explored youth subcultures and nightclub spaces in Sheffield, which involved a thoroughly enjoyable research process frequenting regular club nights! I loved gaining a real insight into a fascinating subject matter, as well as finally being able to get really stuck into areas of theory and literature. As it happens, I was quite disappointed with the mark for my undergraduate dissertation. However, for me, its real value was the way it galvanised my interest in geography and conducting geographical research. Indeed, from these unassuming beginnings of knocking about the pubs and clubs of Sheffield, my career as a geographer has led me to conduct research in Kenya, Indonesia and Sri Lanka through a Masters in International Development and, subsequently, a PhD in Human Geography. The latter involved nine months of living, working and researching in Sri Lanka as I explored the long term legacies of the 2004 Indian Ocean tsunami.

So what has been gained from undertaking these geographical research projects? On a personal level it has equipped me with some very useful life skills: improved organisation; good time management; an ability to critically interpret, understand and synthesise data; experience of living and working abroad; the list could go on. Conducting geographical research has really allowed me to 'get under the skin' of the locations I have studied. Spending extended periods of time intensely studying places has been a fascinating process, allowing me to encounter multiple worldviews and opinions, not to mention being the catalyst for some wonderful, lasting friendships with people I have met along the way.

Doing geographical research has also helped me to understand my place in the world. For me, engaging with geographical theory and applying it to 'real life' has been a way of unlocking the complex and dynamic world we live in and a way of beginning to make sense of it. It has been a way of exploring its overwhelming challenges, inequalities and disparities, as well as highlighting moments of hope, resistance and positive change. Indeed, in the years I've been 'doing geography' my worldview has shifted radically and, while I still have so many unanswered questions, conducting geographical research has made me feel better equipped to comprehend and confront the multiple challenges that the world faces. Importantly, my research has not been all about me but also had benefits for some of the people I have worked with. As well as numerous positive encounters during the research processes, my Master's research has been used by a local NGO to

(Continued)

(Continued)

inform their work. As I disseminate my PhD research, I am hoping it will have a similar positive impact. This is not only important to me to ensure that my research is not a completely extractive process, but seeing my work have a tangible impact is one of the most rewarding aspects of undertaking these projects.

Personally, the dissertation is often incredibly rewarding, whatever the result. As noted in Chapter 1, the reasons for undertaking an extended project are manifold – but include developing transferable skills that can be carried with you beyond your years at university. For those going into the job market, evidence of a completed dissertation can tell an employer much about your ability to work independently, manage your time, produce a report, and so on. Indeed, many of my own students use feedback from their dissertations in job applications and interviews. For those contemplating further study (a Master's degree or even a PhD) the dissertation is an important stepping-stone to further independent study where you will complete another research project at an advanced level.

More broadly, as geographers we do not just tell stories of our research for ourselves. To return to Chapter 1 once more, the reason we engage with 'earth-writing' is because we are inquisitive; because there are questions (big and small) about the world that require an answer. These answers help us to understand socio-cultural, economic, political and environmental relations between people and place. Accordingly, at the end of your dissertation is useful to ask if the findings of your research should be communicated with others. Often, if your research involved particular case study sites, or engaged with particular groups of people, it is worthwhile asking whether the research you produced would be of use to them. Would they like a copy of your dissertation or a short summary report of your findings? As Walliman notes, this can act as both a 'thank you' to those you worked with, as well as potentially benefiting those people, places and organisations (2014: 265).

Moreover, some students find that maintaining links with people they have worked with during dissertation research can lead to employment opportunities (depending on the nature of the project). Furthermore, if your work has particular academic merit (making intellectual or theoretical advancements) your supervisor may recommend that you consider submitting it for a prize. Research Groups of the Royal Geographical Society (RGS), for example, often run annual competitions for the strongest dissertations authored by undergraduates. A list of groups and prizes available can be found online, on the RGS and individual Research Group webpages. Additionally, if your work has some 'surprising results', 'serious or wide significance' or attends to a topical issue, Walliman notes you could consider whether the work could be published in the press or online (2014: 266).

This is not easy and – given you represent the institution in which you study – it is normally recommended that you discuss any potential to do this with your supervisor or university press office before proceeding.

So now we are at the end. If you are closing the pages of this book it probably means you are nearing the completion of your own human geography dissertation. Certainly, by this point, you will know what geography is, and what geographers do. You will know it is the job of a geographer to write about the world. For the human geographer specifically, they have a job to help make sense of the relations between people, space and place, putting pen to paper in this respect. You will know this because you will have engaged with this at first hand. In designing, doing and delivering your own piece of research you will – like the academic staff in the school or department where you study – have become an 'earth-writer'. You will have produced a piece of knowledge about the world we live in. You will therefore have made a transition from being a *student* of geography to a *producer* of geography. Accordingly then, as the book concludes, I'd like to congratulate you on *being* a geographer.

Chapter Summary

- To finish your human geography dissertation you must insert the bookends – relevant preliminary pages that introduce the project (a title page, declaration, abstract, table of contents and acknowledgements) and concluding pages that finalise it (appendices and a reference list).
- A good dissertation is one that is proof-read and polished. It is also a piece of work that is professionally presented with a carefully considered layout. Images, maps, graphs and tables should all be appropriately formatted and legible, especially when printed. Leave ample time for printing and binding your work before the deadline.
- There is often much mystery around how a dissertation is marked. Consult the criteria to understand what the marker is looking for when they evaluate your work. Most dissertations will be double-marked. When results are released always gain feedback to understand the mark you received – whether good or bad.
- What you learn in completing a dissertation will move with you, providing skills relevant to future employment or continuing study. Use the knowledge gained from your dissertation to enhance job or university applications moving forwards.
- Your dissertation may well hold interest for others. Consider if – and how – you can communicate the findings of your dissertation (via academic channels, the communities or organisations you worked with, in the press or on the web).

Key readings

Bradford, M. (2010) 'Writing essays, reports and dissertations', in N. Clifford, S. French and G. Valentine (eds) *Key Methods in Geography* (second edition). London: Sage. pp. 497–512.

Walliman, N. (2014) *Your Undergraduate Dissertation: The Essential Guide to Success* (second edition). London: Sage (see particularly Chapters 18 and 19).

REFERENCES

Aberystwyth University Ethics Guidance (2014) [unpublished].

Adey, P. (2009) *Mobility*. London: Routledge.

Adey, P. and Anderson, B. (2012) 'Anticipating emergencies: Technologies of preparedness and the matter of security', *Security Dialogue* 43 (2): 99–117.

Adey, P., Whitehead, M. and Williams, A. (eds) (2013) *From Above: War, Violence and Verticality*. London: Hurst & Co.

Agar, M.H. (1980) *The Professional Stranger: An Informal Introduction to Ethnography*. New York: Academic Press.

Agnew, J. (1996) 'Mapping politics: How context counts in electoral geography', *Political Geography* 15 (2): 129–46.

Agnew, J. and Duncan, J. (1989) *The Power of Place*. London: Unwin Hyman.

Aitken, S. (2005) 'Textual analysis: Reading culture and context', in R. Flowerdew and D. Martin (eds) *Methods in Human Geography: A Guide for Students Doing a Research Project* (second edition). London and New York: Routledge. pp. 233–49.

Aitken, S. and Valentine, G. (eds) (2006) *Approaches to Human Geography*. London: Sage.

Anderson, B. and Adey, P. (2011) 'Affect and security: Exercising emergency in "UK civil contingencies"', *Environment and Planning D: Society and Space* 29 (6): 1092–1109.

Anderson, B. and Wylie, J. (2009) 'On geography and materiality', *Environment and Planning A* 41 (2): 318–35.

Anderson, J. (2004) 'Spatial politics in practice: The style and substance of environmental direct action', *Antipode* 36 (1): 106–25.

Anderson, J. (2014) 'What I talk about when I talk about kayaking', in J. Anderson and K. Peters (eds) *Water Worlds: Human Geographies of the Ocean*. Farnham: Ashgate. pp. 103–18.

Anderson, J. (2015) *Understanding Cultural Geography: Places and Traces* (second edition). London and New York: Routledge.

Anderson, J., Adey, P. and Bevan, P. (2010) 'Positioning place: Polylogic approaches to research methodology', *Qualitative Research* 10 (5): 589–604.

Ash, J. and Simpson, P. (2016) 'Geography and post-phenomenology', *Progress in Human Geography* 40 (1): 48–66.

Aswani, S. and Lauer, M. (2006) 'Incorporating fishermen's local knowledge and behavior into geographical information systems (GIS) for designing marine protected areas in Oceania', *Human Organization* 65 (1): 81–102.

Atkinson, D. and Shakespeare, P. (1993) *Reflecting on Research Practice: Issues in Health and Social Welfare*. Buckingham: Open University Press.

Atkinson, R. (2007) 'Ecology of sound: The sonic order of urban space', *Urban Studies* 44 (10): 1905-17.

Baker, A. (1997) '"The dead don't answer questionnaires": Research and writing historical geography', *Journal of Geography in Higher Education* 21: 231-43.

Barnes, T.J. and Duncan, J.S. (1992) *Writing Worlds: Discourse, Text and Metaphor in the Representation of Landscape*. London and New York: Routledge.

Batty, M. (2010) 'Using Geographical Information Systems', in N. Clifford, S. French and G. Valentine (eds) *Key Methods in Geography* (second edition). London: Sage. pp. 408-22.

Bear, C. and Eden, S. (2008) 'Making space for fish: The regional, network and fluid spaces of fisheries certification', *Social and Cultural Geography* 9 (5): 487-504.

Bennett, J. (2004) 'The force of things: Steps toward an ecology of matter', *Political Theory*, 32 (3): 347-72.

Bennett, K., Ekinsmyth, C. and Shurmer-Smith, P. (2002) 'Selecting topics for study', in P. Shurmer-Smith (ed.) *Doing Cultural Geography*. London: Sage. pp. 81-94.

Berg, B.L. (2004) *Qualitative Research Methods for the Social Sciences*. London: Pearson Education.

Billig, M. (1995). *Banal Nationalism*. London: Sage.

Birtchnell, T., Savitzky S. and Urry, J. (eds) (2015) *Cargomobilities: Moving Materials in a Global Age*. London: Routledge.

Bissell, D. (2014) 'Encountering stressed bodies: Slow creep transformations and tipping points of commuting mobilities', *Geoforum* 51: 191-201.

Bissell, D. (2016) 'Micropolitics of mobility: Public transport commuting and everyday encounters with forces of enablement and constraint', *Annals of the American Association of Geographers* 106 (2): 394-403.

Blaut, J.M. (1961) 'Space and process', *The Professional Geographer* 13 (4): 1-7.

Blomley, N. and Delaney, D. (eds) (2001) *Legal Geographies Reader*. Oxford: Blackwell.

Blunt, A. and Wills, J. (2000) *Dissident Geographies: An Introduction to Radical Ideas and Practice*. Harlow: Pearson Education.

Bochove, M. and Engbersen, G. (2015) 'Beyond cosmopolitanism and expat bubbles: Challenging dominant representations of knowledge workers and trailing spouses', *Population, Space and Place* 21 (4): 295-309.

Boland, P. (2010) 'Sonic geography, place and race in the formation of local identity: Liverpool and Scousers', *Geografiska Annaler: Series B, Human Geography* 92 (1): 1–22.

Bondi, L. (2009) 'Teaching reflexivity: Undoing or reinscribing habits of gender?' *Journal of Geography in Higher Education* 33 (3): 327-37.

Bonnett, A. (2014) *How to Argue*. London: Pearson Education.

Bosco, F. and Herman, T. (2010) 'Focus groups as collaborative research performances' in D. DeLyser, S. Herbert, S. Aitken, M. Crang, and L. McDowell (eds) *The Sage Handbook of Qualitative Geography*. London: Sage. pp. 193–207.

Bradford, M. (2010) 'Writing essays, reports and dissertations', in N. Clifford, S. French and G. Valentine (eds) *Key Methods in Geography* (second edition). London: Sage. pp. 497-512.

Braun, B. and Castree, N. (eds) (2005) *Remaking Reality: Nature at the Millennium*. London: Routledge.

Breau, S. (2014) 'The Occupy Movement and the top 1% in Canada', *Antipode*, 46 (1): 13-33.

Breen, R. (2006) 'A practical guide to focus-group research', *Journal of Geography in Higher Education* 30 (3): 463-75.

Breitbart, M. (2010) 'Participatory research methods', in N. Clifford, S. French and G. Valentine (eds) *Key Methods in Geography* (second edition). London: Sage. pp. 141-56.

Brewer, J. (2000) *Ethnography*. Buckingham: Open University Press.

Brickell, K. and Datta, A. (eds) (2011) *Translocal Geographies*. Farnham: Ashgate.

Bridge, G. and Le Billon, P. (2013) *Oil*. Cambridge: Polity.

Bryman, A. (2004) *Social Research Methods* (second edition). Oxford: Oxford University Press.

Bryman, A. (2006) 'Integrating quantitative and qualitative research: How is it done?' *Qualitative Research* 6 (1): 97-113.

Bullard, J. (2010) 'Health and safety in the field', in N. Clifford, S. French and G. Valentine (eds) *Key Methods in Geography* (second edition). London: Sage. pp. 49-58.

Buller, H. (2013) 'Animal geographies I', *Progress in Human Geography* 38 (2): 308-18.

Burkill, S. and Burley, J. (1996) 'Getting started on a geography dissertation', *Journal of Geography in Higher Education* 20 (3): 431-7.

Burrell, K. and Hörschelmann, K. (eds) (2014) *Mobilities in Socialist and Post-Socialist States*. London: Palgrave Macmillan.

Büscher, M. and Urry, J. (2009) 'Mobile methods and the empirical', *European Journal of Social Theory* 12 (1): 99-116.

Butler, J. (1990) *Gender Trouble and the Subversion of Identity*. New York and London: Routledge.

Butler, T. (2006) 'A walk of art: The potential of the sound walk as practice in cultural geography', *Social and Cultural Geography* 7 (6): 889-908.

Butler, T. (2007) 'Memoryscape: How audio walks can deepen our sense of place by integrating art, oral history and cultural geography', *Geography Compass* 1 (3): 360-72.

Buttimer, A. (1976) 'Grasping the dynamism of the lifeworld', *Annals of the Association of American Geographers* 66 (2): 277-92.

Castree, N. and Braun, B. (eds) (2001) *Social Nature: Theory, Practice, and Politics*. Oxford: Blackwell.

Chatterton, P. (2002) '"Squatting is still legal, necessary and free": A brief intervention in the corporate city', *Antipode* 34 (1): 1-7.

Chorley, R. and Haggett, P. (eds) (1967) *Models in Geography*. London: Methuen.

Christaller, W. (1933) *Central Places in Southern Germany*. Jena: Fischer.

Clark, N. (2010) *Inhuman Nature: Sociable Life on a Dynamic Planet*. London: Sage.

Clifford, N., French, S. and Valentine, G. (eds) (2010) *Key Methods in Geography* (second edition). London: Sage.

Clifford, N., Cope, M., Gillespie, T. and French, S. (eds) (2016) *Key Methods in Geography* (third edition). London: Sage.

Clifford, N., French, S. and Valentine, G. (2010) 'Getting started in geographical research', in N. Clifford, S. French and G.Valentine (eds) *Key Methods in Geography* (second edition). London: Sage. pp. 3–15.

Cloke, P., Cook, I., Crang, P., Goodwin, M., Painter, J. and Philo, C. (2004) 'Changing practices in human geography: An introduction', in P. Cloke, I. Cook, P. Crang, M. Goodwin, J. Painter and C. Philo (eds) *Practising Human Geography*. London: Sage. pp. 1-34.

Collins, R. (2005) 'Whitbread Writer', *BBC Wales.* http://downloads.bbc.co.uk/wales/archive/bbc-mid-wales-books-richard-collins-interview.pdf [accessed 2 July 2016].

Colls, R. and Evans, B. (2014) 'Making space for fat bodies? A critical account of "the obesogenic environment"', *Progress in Human Geography* 38 (6): 733–53.

Cook, I. (2001) 'You want to be careful you don't end up like Ian. He's all over the place: Autobiography in/of an expanded field', in P. Moss (ed.) *Placing Autobiography in Geography*. Syracuse: Syracuse University Press. pp. 99–120.

Cook, I. (2004) 'Follow the thing: Papaya', *Antipode* 36 (4): 642–64.

Cope, M. (2010) 'Coding transcripts and diaries', in N. Clifford, S. French and G. Valentine (eds) *Key Methods in Geography* (second edition). London: Sage. pp. 440–52.

Cope, M. and Elwood, S. (eds) (2009) *Qualitative GIS: A Mixed Methods Approach*. London: Sage.

Couper, P. (2014) *A Student's Introduction to Geographic Thought: Theories, Philosophies, Methodologies*. London: Sage.

Crang, M. (1994) 'Spacing times, telling times and narrating the past', *Time & Society* 3 (1): 29–45.

Crang, M. (2005) 'Analysing qualitative materials', in R. Flowerdew and D. Martin (eds) *Methods in Human Geography: A Guide for Students Doing a Research Project* (second edition). London and New York: Routledge. pp. 218–32.

Crang, P. (1994) 'It's showtime: On the workplace geographies of display in a restaurant in southeast England', *Environment and Planning D: Society and Space* 12 (6): 675–704.

Creative Commons (2015) Hourglass image (n.d.) VistaICO.com (Creative Commons via Wikimedia Commons) https://upload.wikimedia.org/wikipedia/commons/2/2a/1328101886_HourGlass.png [accessed 17 June 2016].

Creswell, J.W. (2013). *Research Design: Qualitative, Quantitative, and Mixed Methods Approaches* (fourth edition). London: Sage.

Cresswell, T. (1996) *In Place/Out of Place: Geography, Ideology, and Transgression*. Minnesota: University of Minnesota Press.

Cresswell, T. (2004) *Place: A Short Introduction*. Oxford: Blackwell.

Cresswell, T. (2006) *On the Move.* London and New York: Routledge.

Cresswell, T. (2013) *Geographic Thought: A Critical Introduction*. London: Wiley Blackwell.

Cresswell, T. and Merriman, P. (eds) (2012) *Geographies of Mobilities: Practices, Spaces, Subjects*. Farnham: Ashgate.

Crouch, D. (2002) 'Surrounded by place: Embodied encounters', in S. Coleman and M. Crang (eds) *Tourism: Between Place and Performance.* Oxford: Berghahn Books. pp. 207–18.

Davies, A. (2013) 'Identity and the assemblages of protest: The spatial politics of the Royal Indian Navy Mutiny, 1946', *Geoforum* 48: 24-32.

Dear, M. (1988) 'The post-modern challenge: Reconstructing human geography', *Transactions of the Institute of British Geographers* 13: 262-74.

Deleuze, G. and Guattari, F. (1988) *A Thousand Plateaus*. London: Athlone Press.

DeLyser, D. (2003) 'Teaching graduate students to write: A seminar for thesis and dissertation writers', *Journal of Geography in Higher Education* 27 (2): 169-81.

DeLyser, D. (2010) 'Writing it up', in B. Gomez and J.P. Jones III (eds) *Research Methods in Geography: A Critical Introduction*. Oxford: Blackwell. pp. 424-36.

DeLyser, D., Herbert, S., Aitken, S., Crang, M. and McDowell, L. (eds) (2010) *The Sage Handbook of Qualitative Geography*. London: Sage.

DeLyser, D. and Sui, D. (2013) 'Crossing the qualitative–quantitative divide II: Inventive approaches to big data, mobile methods, and rhythmanalysis', *Progress in Human Geography* 37 (2): 293-305.

DeLyser, D. and Sui, D. (2014) 'Crossing the qualitative–quantitative chasm III: Enduring methods, open geography, participatory research, and the fourth paradigm', *Progress in Human Geography* 38 (2): 294-307.

Denscombe, M. (2007) *The Good Research Guide: For Small-Scale Social Research Projects.* Maidenhead: McGraw Hill Education, Open University Press.

Dhanju, R. and O'Reilly, K. (2013) 'Human subjects research and the ethics of intervention: Life, death, and radical geography in practice', *Antipode* 45 (3): 513-16.

Dittmer, J. (2005) 'Captain America's empire: Reflections on identity, popular culture, and post-9/11 geopolitics', *Annals of the Association of American Geographers* 95 (3): 626-43.

Dittmer, J. (2012) *Captain America and the Nationalist Superhero: Metaphors, Narratives, and Geopolitics.* Philadelphia: Temple University Press.

Dixon, D. and Straughan, E. (2010) 'Geographies of touch/touched by geography', *Geography Compass* 4 (5): 449-59.

Dodds, K. (2006) 'Popular geopolitics and audience dispositions: James Bond and the internet movie database (IMDb)', *Transactions of the Institute of British Geographers* 31 (2): 116-30.

Doel, M. (1996) 'A hundred thousand lines of flight: A machinic introduction to the nomad thought and scrumpled geography of Gilles Deleuze and Félix Guattari', *Environment and Planning D: Society and Space* 14 (4): 421-39.

Doel, M. (2010) 'Analysing cultural texts', in N. Clifford, S. French and G. Valentine (eds) *Key Methods in Geography* (second edition). London: Sage. pp. 485-96.

Domosh, M. (1991) 'Towards a feminist historiography of geography', *Transactions of the Institute of British Geographers* 16: 95-104.

Dorling, D. (2010) 'Using statistics to describe and explore data', in N. Clifford, S. French and G. Valentine (eds) *Key Methods in Geography* (second edition). London: Sage. pp. 374-85.

Driver, F. (2003) 'On geography as a visual discipline', *Antipode* 35 (2): 227-31.

Dummer, T., Cook, I., Parker, S., Barrett, G. and Hull, A. (2008) 'Promoting and assessing "deep learning" in geography fieldwork: An evaluation of reflective field diaries', *Journal of Geography in Higher Education* 32 (3): 459-79.

Dwyer, C. and Davies, G. (2010) 'Qualitative methods III: Animating archives, artful interventions and online environments', *Progress in Human Geography* 34 (1): 88-97.

Economic and Social Research Council (ESRC) (2016) 'Our policy and guidelines for good research conduct'. http://www.esrc.ac.uk/funding/guidance-for-applicants/research-ethics/our-policy-and-guidelines-for-good-research-conduct/ [accessed 3 July 2016].

Edensor, T. (2000) 'Walking in the British countryside: Reflexivity, embodied practices and ways to escape', *Body & Society* 6 (3-4): 81-106.

Edensor, T. (2005) 'The ghosts of industrial ruins: Ordering and disordering memory in excessive space', *Environment and Planning D: Society and Space* 23 (6): 829-49.

Edensor, T. (2007) 'Sensing the ruin', *The Senses and Society* 2 (2): 217-32.

Edensor, T. (2010) 'Walking in rhythms: Place, regulation, style and the flow of experience', *Visual Studies* 25 (1): 69-79.

Edensor, T. (2013) 'Reconnecting with darkness: Gloomy landscapes, lightless places', *Social and Cultural Geography* 14 (4): 446-65.

Elden, S. (2013) 'Secure the volume: Vertical geopolitics and the depth of power', *Political Geography* 34: 35-51.

Elwood, S. (2010) 'Mixed methods: Thinking, doing, and asking in multiple ways', in D. DeLyser, S. Herbert, S. Aitken, M. Crang and L. McDowell (eds) *Sage Handbook of Qualitative Geography*. London: Sage. pp. 94-113.

Elwood, S. and Mitchell, K. (2012) 'Mapping children's politics: Spatial stories, dialogic relations and political formation', *Geografiska Annaler: Series B, Human Geography* 94 (1): 1-15.

England, K. (1994) 'Getting personal: Reflexivity, positionality, and feminist research', *The Professional Geographer* 46 (1): 80-9.

Entriken, N.J. (1976) 'Contemporary humanism in geography', *Annals of the Association of American Geographers* 66 (4): 615-32.

ESRI (n.d.) 'What is GIS?' http://www.esri.com/what-is-gis [accessed 28 June 2016].

Evers, C. (2009) '"The Point": Surfing, geography and a sensual life of men and masculinity on the Gold Coast, Australia', *Social and Cultural Geography* 10 (8): 893-908.

Feminist Pedagogy Working Group (2002) *Defining Feminism?* London: Royal Geographical Society.

Field, R. (2010) 'Data handling and representation', in N. Clifford, S. French and G. Valentine (eds) *Key Methods in Geography* (second edition). London: Sage. pp. 317-49.

Fincham, B., McGuiness, M. and Murray, L. (2010) *Mobile Methodologies*. London: Palgrave Macmillan.

Flowerdew, R. (2005) 'Finding previous work on the topic', in R. Flowerdew and D. Martin (eds) *Methods in Human Geography: A Guide for Students Doing a Research Project* (second edition). London and New York: Routledge. pp. 48-56.

Flowerdew, R. and Martin, D. (eds) (1997) *Methods in Human Geography: A Guide for Students Doing a Research Project*. Harlow: Longman.

Forsyth, I. (2013) 'Subversive patterning: The surficial qualities of camouflage', *Environment and Planning A* 45 (5): 1037-52.

Foster, J. and Sheppard, J. (2002) *British Archives: A Guide to Archival Resources in the UK*. Basingstoke: Macmillan.

Fotheringham, S., Brunsdon, C. and Charlton, M. (2007) *Quantitative Geography: Perspectives on Spatial Data Analysis*. London: Sage.

Foucault, M. (2013) *Archaeology of Knowledge*. London and New York: Routledge.

Foucault, M. and Miskowiec, J. (1986) 'Of other spaces', *Diacritics* 16 (1): 22-7.

Garrett, B.L. (2011a) 'Cracking the Paris carrières: Corporal terror and illicit encounter under the city of light', *ACME: An International E-Journal for Critical Geographies* 10 (2): 269-77.

Garrett, B.L. (2011b) 'Videographic geographies: Using digital video for geographic research', *Progress in Human Geography* 35 (4): 521-41.

Garrett, B.L. (2014) 'Undertaking recreational trespass: Urban exploration and infiltration', *Transactions of the Institute of British Geographers* 39 (1): 1-13.

Garrett, B.L. and Brickell, K. (2015) 'Participatory politics of partnership: Video workshops on domestic violence in Cambodia', *Area* 47 (3): 230-6.

Gatrell, A.C (1997) 'Choosing a topic', in R. Flowerdew and D. Martin (eds) *Methods in Human Geography: A Guide for Students Doing a Research Project*. Harlow: Longman. pp. 36-45.

Gatrell, A. and Flowerdew, R. (2005) 'Choosing a topic', in R. Flowerdew and D. Martin (eds) *Methods in Human Geography: A Guide for Students Doing a Research Project* (second edition). London and New York: Routledge. pp. 38-47.

Geoghegan, H. (n.d.) 'The culture of enthusiasm: Citizen science' tag: https://hilarygeoghegan.wordpress.com/tag/citizen-science/ [accessed 29 June 2016].

Geoghegan, H. (2010) 'Museum geography: Exploring museums, collections and museum practice in the UK', *Geography Compass* 4 (10): 1462-76.

Goetz, A.R., Vowles, T.M. and Tierney, S. (2009) 'Bridging the qualitative–quantitative divide in transport geography', *The Professional Geographer* 61 (3): 323-35.

Goodchild, M. (2010) 'Geographic Information Systems', in B. Gomez and J.P. Jones III (eds) *Research Methods in Geography: A Critical Introduction*. Oxford: Blackwell. pp. 376-91.

Graham, E. (2005) 'Philosophies underlying human geography research', in R. Flowerdew and D. Martin (eds) *Methods in Human Geography: A Guide for Students Doing a Research Project* (second edition). London and New York: Routledge. pp. 8-33.

Gregory, D. (1978) *Ideology, Science and Human Geography*. London: Hutchinson & Co.

Gregory, D., Johnston, R., Pratt, G., Watts, M. and Whatmore, S. (eds) (2011) *The Dictionary of Human Geography*. Oxford: John Wiley & Sons.

Hall, S. (1980) 'Encoding/decoding', in S. Hall, D. Hobson, A. Lowe and P. Willis (eds) *Culture, Media, Language*. London: Unwin Hyman. pp. 117–27.

Hannam, K. (2002) 'Using archives', in P. Shurmer-Smith (ed.) *Doing Cultural Geography*. London: Sage. pp. 113-22.

Haraway, D. (1988) 'Situated knowledges: The science question in feminism and the privilege of partial perspective', *Feminist Studies* 14 (3): 575-99.

Haraway, D. (1991) *Simians, Cyborgs, and Women: The Reinvention of Nature*. New York: Routledge.

Harding, S .(1986) *The Science Question in Feminism*. Milton Keynes: Open University Press.

Hardy, A., Mageni, Z., Dongus, S., Killeen, G., Macklin, M., Majambare, S., Ali, A., Msellem, M., Al-Mafazy, A.-W., Smith, M. and Thomas, C. (2015) 'Mapping hotspots of malaria transmission from pre-existing hydrology, geology and geomorphology data in the pre-elimination context of Zanzibar, United Republic of Tanzania', *Parasites and Vectors* 8 (1): 1–15.

Hart, C. (1999) *Doing a Literature Review: Releasing the Social Science Imagination*. London: Sage.

Harvey, D. (1969) *Explanation in Geography*. London: Edward Arnold.

Harvey, D. (1973) *Social Justice and the City*. London: Edward Arnold.

Harvey, D. (1989) *The Condition of Postmodernity*. Oxford: Blackwell.

Hasty, W. (2011) 'Piracy and the production of knowledge in the travels of William Dampier, c. 1679–1688', *Journal of Historical Geography* 37 (1): 40-54.

Hawkins, H. (2015) 'Creative geographic methods: Knowing, representing, intervening. On composing place and page', *Cultural Geographies* 22 (2): 247-68.

Hay, I. (2010) 'Ethical practice in geographic research', in N. Clifford, S. French and G. Valentine (eds) *Key Methods in Geography* (second edition). London: Sage. pp. 35-48.

Hay, I. (2012) *Communicating in Geography and the Environmental Sciences* (fourth edition). Oxford: Oxford University Press.

Heley, J. (2011) 'On the potential of being a village boy: An argument for local rural ethnography', *Sociologia Ruralis* 51 (3): 219-37.

Heley, M. and Heley, R. (2010) 'How to conduct a literature search', in N. Clifford, S. French and G. Valentine (eds) *Key Methods in Geography*, (second edition). London: Sage. pp. 16-34.

Herbert, S. and Brown, E. (2006) 'Conceptions of space and crime in the punitive neoliberal city', *Antipode* 38 (4): 755-77.

Hitchings, R. (2003) 'People, plants and performance: On actor network theory and the material pleasures of the private garden', *Social and Cultural Geography* 4 (1): 99-114.

Hoby, H. (2013) 'Margaret Atwood: Interview', http://www.telegraph.co.uk/culture/books/10246937/Margaret-Atwood-interview.html [accessed 2 July 2016].

Hoggart, K., Lees, L. and Davies, A. (2002) *Researching Human Geography*. London: Arnold.

Holloway, L. and Hubbard, P. (2001). *People and Place: The Extraordinary Geographies of Everyday Life*. Harlow: Pearson Education.

Holton, M. and Riley, M. (2014) 'Talking on the move: Place-based interviewing with undergraduate students', *Area* 46 (1): 59-65.

hooks, b. (1990) *Ain't I a Woman: Black Women and Feminism*. London: Pluto Press.

Hopkins, P. and Noble, G. (2009) 'Masculinities in place: Situated identities, relations and intersectionality', *Social and Cultural Geography* 10 (8): 811–19.

Horton, J. and Kraftl, P. (2006) 'What else? Some more ways of thinking and doing "children's geographies"', *Children's Geographies* 4 (1): 69–95.

Hoskins, G. (2007) 'Materialising memory at Angel Island Immigration Station, San Francisco', *Environment and Planning A* 39: 437–55.

Hyde, A. (2015) 'Inhabiting no-man's-land: The military mobilities of army wives' (Doctoral dissertation, London School of Economics and Political Science (LSE)). Available from http://etheses.lse.ac.uk/3142/ [accessed 24 June 2016].

Hyndman, J. (2004) 'Mind the gap: Bridging feminist and political geography through geopolitics', *Political Geography* 23 (3): 307–22.

Ingold, T. (2004) 'Culture on the ground: The world perceived through the feet', *Journal of Material Culture* 9 (3): 315–40.

Ingold, T. (2007) *Lines: A Brief History.* London and New York: Routledge.

Ingold, T. (2011a) *Being Alive: Essays on Movement, Knowledge and Description.* London and New York: Routledge.

Ingold, T. (2011b) *The Perception of the Environment: Essays on Livelihood, Dwelling and Skill.* London and New York: Routledge.

Jackson, L. (2011) 'Mixed methodologies in emotive research: Negotiating multiple methods and creating narratives in feminist embodied work on citizenship', *Graduate Journal of Asia Pacific Studies* 7 (2): 46–61.

Jackson, L. and Valentine, G. (2016) 'Rethinking spaces, sites and encounters of conflict in twenty-first century Britain: The case of abortion protest in public space', in M. De Backer, L. Melcago, G. Varna and F. Menichelli (eds) *Order and Conflict in Public Space.* Abingdon: Routledge. pp. 182–205.

Jackson, P. (2000) 'Rematerializing social and cultural geography', *Social and Cultural Geography* 1 (1): 9–14.

Jazeel, T. (2014) 'Subaltern geographies: Geographical knowledge and postcolonial strategy', *Singapore Journal of Tropical Geography* 35 (1): 88–103.

Johnston, L. (1996) 'Flexing femininity: Female body-builders refiguring "the body"', *Gender, Place and Culture* 3 (3): 327–40.

Johnston, R. (1986) *Philosophy and Human Geography: An Introduction to Contemporary Approaches.* London: Arnold.

Johnston, R. and Pattie, C. (2011) 'The British general election of 2010: A three-party contest – or three two-party contests?' *The Geographical Journal* 177 (1): 17–26.

Johnston, R. and Sidaway, J. (2004) *Geography and Geographers: Anglo-American Human Geography Since 1945.* London: Arnold.

Kanngieser, A. (2012) 'A sonic geography of voice: Towards an affective politics', *Progress in Human Geography* 36 (3): 336–53.

Kindon, S., Pain, R. and Kesby, M. (eds) (2007) *Participatory Action Research Approaches and Methods: Connecting People, Participation and Place.* Abingdon: Routledge.

Kinsley, S. (2014) 'The matter of "virtual" geographies', *Progress in Human Geography*, 38 (3): 364–84.

Kitchin, R. (1998) 'Towards geographies of cyberspace', *Progress in Human Geography* 22 (3): 385-406.

Kitchin, R. (2013) 'Big data and human geography: Opportunities, challenges and risks', *Dialogues in Human Geography* 3 (3): 262-7.

Kitchin, R. and Tate, N. (2000) *Conducting Research in Human Geography: Theory, Methodology and Practice*. Harlow: Prentice Hall.

Kneale, P. (2003) *Study Skills for Geography Students: A Practical Guide* (second edition). London: Arnold.

Kneale, P. (2011) *Study Skills for Geography, Earth and Environmental Science Students* (third edition). Abingdon: Hodder Education.

Kozinets, R. (2002) 'The field behind the screen: Using netnography for marketing research in online communities', *Journal of Marketing Research* 39 (1): 61-72.

Kullman, K. (2013) 'Geographies of experiment/experimental geographies: A rough guide', *Geography Compass* 7 (12): 879-94.

Kwan, M.P. (2002) 'Feminist visualization: Re-envisioning GIS as a method in feminist geographic research', *Annals of the Association of American Geographers* 92 (4): 645-61.

Kwan, M.P. and Ding, G. (2008) 'Geo-narrative: Extending geographic information systems for narrative analysis in qualitative and mixed-method research', *The Professional Geographer* 60 (4): 443-65.

Lamott, A. (1995) *Bird by Bird: Some Instructions on Writing and Life*. New York: Anchor Books.

Last, A. (2012) 'Experimental geographies', *Geography Compass* 6 (12): 706-24.

Laurier, E. (2010) 'Participant observation', in N. Clifford, S. French and G. Valentine (eds) *Key Methods in Geography* (second edition). London: Sage. pp. 116-30.

Law, J. and Urry, J. (2004) 'Enacting the social', *Economy and Society* 33 (3): 390-410.

Le Billon, P. (2008) 'Diamond wars? Conflict diamonds and geographies of resource wars', *Annals of the Association of American Geographers* 98 (2): 345-72.

Leach, E. (1961) *Rethinking Anthropology*. London: Athlone Press.

Lefebvre, H. (1991) *The Production of Space*. Oxford: Blackwell.

Limerick, P.N. (2012) 'Dancing with professors: The trouble with academic prose', in V. Zamel and R. Spack (eds) *Negotiating Academic Literacies*. London: Routledge. pp. 199–206.

Longhurst, R. (1997) '(Dis)embodied geographies', *Progress in Human Geography* 21 (4): 486-501.

Longhurst, R. (2006) 'Plots, plants and paradoxes: contemporary domestic gardens in Aotearoa/New Zealand', *Social and Cultural Geography* 7 (4): 581-93.

Longhurst, R. (2010) 'Semi-structured interviews and focus groups', in N. Clifford, S. French and G. Valentine (eds) *Key Methods in Geography* (second edition). London: Sage. pp.103-15.

Longley, P., Cheshire, J. and Mateos, P. (2011) 'Creating a regional geography of Britain through the spatial analysis of surnames', *Geoforum* 42 (4): 506-16.

Lorimer, H. (2005) 'Cultural geography: The busyness of being "more-than-representational"', *Progress in Human Geography* 29 (1): 83-94.

Lynch, M. (2000) 'Against reflexivity as an academic virtue and source of privileged knowledge', *Theory, Culture & Society*, 17 (3): 26-54.
Madge, C. (2010) 'Internet mediated research', in N. Clifford, S. French and G. Valentine (eds) *Key Methods in Geography* (second edition). London: Sage. pp. 173-88.
Madge, C. (2014) 'On the creative (re) turn to geography: Poetry, politics and passion', *Area* 46 (2): 178-85.
Madge, C. and O'Connor, H. (2002) 'On-line with e-mums: Exploring the internet as a medium for research', *Area* 34 (1): 92-102.
Madge, C. and O'Connor, H. (2006) 'Parenting gone wired: Empowerment of new mothers on the internet?', *Social and Cultural Geography* 7 (2): 199-220.
Marx, K. (1867) Capital: Volume 1: Section 4: The Fetishism of Commodities and the Secret Thereof. https://www.marxists.org/archive/marx/works/1867-c1/ch01.htm [accessed 3 July 2016].
Massey, D. (1984) *Spatial Divisions of Labour: Social Structures and the Geography of Production*. Basingstoke and London: Macmillan.
Massey, D. (1997) 'A global sense of place', in D. Gregory and T. Barnes (eds) *Reading Human Geography: The Poetics and Politics of Enquiry*. London: Arnold. pp. 315-23.
Massey, D. (2005) *For Space*. London: Sage.
Matless, D. (2005) 'Sonic geography in a nature region', *Social and Cultural Geography* 6 (5): 745-66.
McDowell, L. (1997) *Capital Culture: Gender at Work in the City*. Oxford: Blackwell.
McDowell, L. (1999) *Gender, Identity and Place: Understanding Feminist Geographies*. Minnesota: University of Minnesota Press.
McLafferty, S.L. (2010) 'Conducting questionnaire surveys', in N. Clifford, S. French and G. Valentine (eds) *Key Methods in Geography* (second edition). London: Sage. pp. 77-88.
Mels, T. (ed.) (2004) *Reanimating Places: A Geography of Rhythms*. Farnham: Ashgate.
Merchant, S. (2011) 'The body and the senses: Visual methods, videography and the submarine sensorium', *Body and Society* 17 (1): 53-72.
Merriman, P. (2012) 'Human geography without time-space', *Transactions of the Institute of British Geographers* 37 (1): 13-27.
Merriman, P. (2014) 'Rethinking mobile methods', *Mobilities* 9 (2): 167-87.
Middleton, J. (2011) 'Walking in the city: The geographies of everyday pedestrian practices', *Geography Compass* 5 (2): 90-105.
Middleton, J. and Yarwood, R. (2015) '"Christians, out here?" Encountering street-pastors in the post-secular spaces of the UK's night-time economy', *Urban Studies* 52 (3): 501-16.
Mills, S. (2013) 'Surprise! Public historical geographies, user engagement and voluntarism', *Area* 45 (1): 16-22.
Mitchell, D. (2000) *Cultural Geography: A Critical Introduction*. Oxford: Blackwell.
Mkono, M. (2011) 'The othering of food in touristic eatertainment: A netnography', *Tourist Studies* 11 (3): 253-70.

Moss, P. (1992) '"Where is the 'Promised Land?": Class and gender in Bruce Springsteen's rock lyrics', *Geografiska Annaler. Series B. Human Geography*, 74 (3): 167-87.

Moss, P. (1995) 'Embeddedness in practice, numbers in context: The politics of knowing and doing', *The Professional Geographer* 47: 442-9.

Moss, P. (ed.) (2002) *Feminist Geography in Practice*. Oxford: Blackwell.

Moss, P. (2011) 'Still searching for the Promised Land: Placing women in Bruce Springsteen's lyrical landscapes', *Cultural Geographies* 18 (3): 343-62.

Murray, R. (2007) *How to Write a Thesis* (second edition). Maidenhead: McGraw-Hill Education.

Nansen, F. (1897) *Farthest North: The Voyage of the 'Fram' 1893–96 and the Fifteen Months Sledge Expedition*. Volume II. Westminster: Constable & Co.

Nash, C. (1996) 'Reclaiming vision: Looking at landscape and the body', *Gender, Place and Culture: A Journal of Feminist Geography*, 3 (2): 149-70.

Nayak, A. and Jeffery, A. (2011) *Geographical Thought: An Introduction to Ideas in Human Geography*. London: Pearson.

Nora, P. (1989) 'Between memory and history: Les lieux de mémoire', *Representations* 26: 7-24.

North, P. (2006) *Alternative Currency Movements as a Challenge to Globalisation? A Case Study of Manchester's Local Currency Networks*. Farnham: Ashgate.

Ogborn, M. (2003) 'Knowledge is power: Using archival research to interpret state formation', in A. Blunt, P. Gruffudd, J. May, M. Ogborn and D. Pinder (eds) *Cultural Geography in Practice*. London: Arnold. pp. 9-22.

Olive, R. (2016) 'Going surfing/doing research: Learning how to negotiate cultural politics from women who surf', *Continuum: Journal of Media & Cultural Studies* 30 (2): 171–82.

Osmond, A. (2016) *Academic Writing and Grammar for Students* (second edition). London: Sage.

Oxford English Dictionary (OED) (2013) 'Oxford English Dictionary', http://www.oed.com [accessed 20/08/2013].

Pain, R. (2004) 'Social geography: Participatory research', *Progress in Human Geography* 28 (5): 652–63.

Pain, R. and Francis, P. (2003) 'Reflections on participatory research', *Area* 35 (1): 46-54.

Panelli, R. (2010) 'More-than-human social geographies: Posthuman and other possibilities', *Progress in Human Geography* 34 (1): 79-87.

Parsons, T. and Knight, P. (2005) *How to Do Your Dissertation in Geography and Related Disciplines* (second edition). London and New York: Routledge.

Paterson, M. (2009) 'Haptic geographies: Ethnography, haptic knowledges and sensuous dispositions', *Progress in Human Geography* 33 (6): 766-88.

Peet, R. (1998) *Modern Geographical Thought*. Oxford: Blackwell.

Perkins, C. (2010) 'Mapping and graphicacy', in N. Clifford, S. French and G. Valentine (eds) *Key Methods in Geography* (second edition). London: Sage. pp. 350-73.

Peters, K. (2011a) 'Sinking the radio "pirates": Exploring British strategies of governance in the North Sea', 1964–1991, *Area* 43 (3): 281-7.

Peters, K. (2011b) 'Negotiating the "place" and "placement" of banal tourist souvenirs in the home', *Tourism Geographies* 13 (2): 234-56.

Peters, K. (2014) 'Material transformations: Place, process and the capacity of tourist souvenirs in the home', in G. Lean, R. Staiff and E. Waterton (eds) *Travel and Transformation*. Farnham: Ashgate. pp. 205-21.

Peters, K. and Turner, J. (2015) 'Between crime and colony: Interrogating (im)mobilities aboard the convict ship', *Social and Cultural Geography* 16 (7): 844–62.

Petersen, J.F., Sack, D. and Garbler, R.E. (2016) *Physical Geography*. Boston: Cengage Learning.

Phillips, R. and Johns, J. (2012) *Fieldwork for Human Geography*. London: Sage.

Philo, C. and Wilbert, C. (2000) *Animal Spaces, Beastly Places: New Geographies of Human–Animal Relations*. London: Routledge.

Pickerill, J. (2003) *Cyberprotest: Environmental Activism Online*. Manchester: Manchester University Press.

Pickerill, J. and Krinsky, J. (2012) 'Why does Occupy matter?' *Social Movement Studies* 11 (3/4): 279-87.

Pink, S. (2012) *Advances in Visual Methodologies*. London: Sage.

Pred, A. (1984) 'Place as historically contingent process: Structuration and the time-geography of becoming places', *Annals of the Association of American Geographers* 74 (2): 279-97.

Pred, A. (1995) *Recognising European Modernities: A Montage of the Present*. London and New York: Routledge.

Price, L. (2015) 'Knitting and the city', *Geography Compass* 9 (2): 81-95.

Pullman, P. (n.d.) 'Author Spotlight', https://www.randomhouse.com/kids/catalog/author.pperl?authorid=24658&view=sml_sptlght [accessed 2 July 2016].

Punch, S. (2012) 'Hidden struggles of fieldwork: Exploring the role and use of field diaries', *Emotion, Space and Society* 5 (2): 86-93.

QAA (2014) 'Subject Benchmarking Statement: Geography.' http://www.qaa.ac.uk/en/Publications/Documents/SBS-geography-14.pdf [accessed 17 June 2016].

Raghuram, P. and Madge, C. (2006) 'Towards a method for postcolonial development geography? Possibilities and challenges', *Singapore Journal of Tropical Geography* 27 (3): 270-88.

Ricketts Hein, J., Evans, J. and Jones, P. (2008) 'Mobile methodologies: Theory, technology and practice', *Geography Compass* 2 (5): 1266-85.

Rodaway, P. (2002) *Sensuous Geographies: Body, Sense and Place*. Abingdon: Routledge.

Rogers, A. (2012) 'Geographies of the performing arts: Landscapes, places and cities', *Geography Compass* 6 (2): 60-75.

Rogers, A. (2015) *Performing Asian Transnationalisms: Theatre, Identity and the Geographies of Performance*. London and New York: Routledge.

Rogerson, P. (2014) *Statistical Methods for Geographers: A Student's Guide* (fourth edition). London: Sage.

Rose, G. (1993) 'Progress in geography and gender. Or something else', *Progress in Human Geography* 17 (4): 531-7.

Rose, G. (1997) 'Situating knowledges: Positionality, reflexivities and other tactics', *Progress in Human Geography* 21 (3): 305-20.

Rose, G. (2003) 'On the need to ask how, exactly, is geography "visual"?', *Antipode* 35 (2): 212-21.

Rose, G. (2012) *Visual Methodologies: An Introduction to Researching with Visual Materials* (third edition). London: Sage.

Rubin, H.J. and Rubin, I.S. (2005) *Qualitative Interviewing: The Art of Hearing Data* (second edition). London: Sage.

Said, E. (1996) *Orientalism*. London: Penguin.

Sayer, A. (1997) 'Essentialism, social constructionism, and beyond', *The Sociological Review* 45 (3): 453-87.

Schutt, R. (2006) *Investigating the Social World*. London: Sage.

Schwartz, J.M. (1996) 'The geography lesson: Photographs and the construction of imaginative geographies', *Journal of Historical Geography* 22 (1): 16-45.

Seltz, S. (2016) 'Pixilated partnerships, overcoming obstacles in qualitative interviews via Skype: A research note', *Qualitative Research* 16 (2): 229-35.

Shaw, W.S., DeLyser, D. and Crang, M. (2015) 'Limited by imagination alone: Research methods in cultural geographies', *Cultural Geographies* 22 (2): 211-15.

Sheller, M. and Urry, J. (2006) 'The new mobilities paradigm', *Environment and Planning A* 38: 207-26.

Shields, R. (ed.) (1996) *Cultures of the Internet: Virtual Spaces, Real Histories, Living Bodies*. London: Sage.

Shurmer-Smith, P. (ed.) (2002) *Doing Cultural Geography*. London: Sage.

Sidaway, J.D. (2000) 'Postcolonial geographies: An exploratory essay', *Progress in Human Geography* 24 (4): 591-612.

Simpson, P. (2011) 'Street performance and the city: Public space, sociality, and intervening in the everyday', *Space and Culture* 14 (4): 415-30.

Smith, D. (2008) 'The politics of studentification and (un)balanced urban populations: Lessons for gentrification and sustainable communities?' *Urban Studies* 45 (12): 2541-64.

Soja, E. (1989) *Postmodern Geographies: The Reassertation of Space in Critical Social Theory*. New York: Verso.

Soja, E. (1995) *Recognizing European Modernities: A Montage of the Present*. London and New York: Routledge.

Spence, E. (2014) 'Unravelling the politics of super-rich mobility: A study of crew and guest on board luxury yachts', *Mobilities* 9 (3): 401-13.

Spinney, J. (2011) 'A chance to catch a breath: Using mobile video ethnography in cycling research', *Mobilities* 6 (2):161-82.

Sporton, D. (1999) 'Mixing methods in fertility research', *The Professional Geographer* 51 (1): 68-76.

Stewart, D. and Shamdasani, P. (2015) *Focus Groups: Theory and Practice*. London: Sage.

Sui, D. and DeLyser, D. (2012) 'Crossing the qualitative–quantitative chasm I: Hybrid geographies, the spatial turn, and volunteered geographic information (VGI)', *Progress in Human Geography* 36 (1): 111-24.

Sultana, F. (2007) 'Reflexivity, positionality and participatory ethics: Negotiating fieldwork dilemmas in international research', *ACME: An International E-Journal for Critical Geographies* 6 (3): 374-85.

Taylor, M. (2014) '"Being useful" after the Ivory Tower: Combining research and activism with the Brixton Pound', *Area* 46 (3): 305-12.

Tuan, Y.F. (1977) *Space and Place: The Perspective of Experience*. Minneapolis: University of Minnesota Press.

Tuan, Y.F. (1991) 'Language and the making of place: A narrative-descriptive approach', *Annals of the Association of American Geographers* 81 (4): 684-96.

Turner, J. and Peters, K. (eds) (2016) *Carceral Mobilities: Interrogating Movement in Incarceration*. London: Routledge.

Urry, J. (2007) *Mobilities*. Cambridge: Polity Press.

Valentine, G. (1996) '(Re)negotiating the "heterosexual street": Lesbian productions of space', in N. Duncan (ed.) *BodySpace: Destablising Geographies of Gender and Sexuality*. New York and London: Routledge. pp. 145-54.

Valentine, G. (1997) 'Tell me about...using interviews as a research methodology', in R. Flowerdew and D. Martin (eds) *Methods in Human Geography: A Guide for Students Doing a Research Project*. Harlow: Longman. pp. 110-26.

Valentine, G. (2003) 'Geography and ethics: In pursuit of social justice ethics and emotions in geographies of health and disability research', *Progress in Human Geography* 27 (3): 375-80.

Valentine, G. (2005) 'Tell me about...using interviews as a research methodology', in R. Flowerdew and D. Martin (eds) *Methods in Human Geography: A Guide for Students Doing a Research Project* (second edition). London and New York: Routledge. pp. 110-27.

van Riper, C.J., Kyle, G.T., Sutton, S.G., Barnes, M. and Sherrouse, B.C. (2012) 'Mapping outdoor recreationists' perceived social values for ecosystem services at Hinchinbrook Island National Park, Australia', *Applied Geography* 35 (1): 164-73.

Waitt, G. (2008) '"Killing waves": surfing, space and gender', *Social and Cultural Geography* 9 (1): 75-94.

Walliman, N. (2011) *Your Research Project: Designing and Planning Your Work* (third edition). London: Sage.

Walliman, N. (2014) *Your Undergraduate Dissertation: The Essential Guide to Success* (second edition). London: Sage.

Ward, K. (2007) 'Geography and public policy: Activist, participatory, and policy geographies', *Progress in Human Geography* 31 (5): 695-705.

Ward, K. (ed.) (2013) *Researching the City*. London: Sage.

Warf, B. (ed.) (2006) *Encyclopedia of Human Geography*. London: Sage.
Weber, A. (1909) *Theory of the Location of Industries*. Chicago: Chicago University Press.
Whatmore, S. (2002) *Hybrid Geographies: Natures, Cultures, Spaces*. London: Sage.
Whatmore, S. (2006) 'Materialist returns: practising cultural geography in and for a more-than-human world', *Cultural Geographies*, 13: 600–9.
Werritty, A. (2006) 'Sustainable flood management: Oxymoron or new paradigm?', *Area* 38 (1): 16–23.
Wheaton, B. (2013) *The Cultural Politics of Lifestyle Sports*. Abingdon: Routledge.
White, P. (2010) 'Making use of secondary data', in N. Clifford, S. French and G. Valentine (eds) *Key Methods in Geography* (second edition). London: Sage. pp. 61–76.
Willis, K. (2009) 'Development: Critical approaches in human geography', in N. Clifford, S. Holloway, S. Rice and G. Valentine (eds) *Key Concepts in Geography* (second edition). London: Sage. pp. 365–77.
Winchester, H.P. (1999) 'Interviews and questionnaires as mixed methods in population geography: The case of lone fathers in Newcastle, Australia', *The Professional Geographer* 51 (1): 60–7.
Withers, C. (2003) 'Constructing "the geographical archive"', *Area* 34 (3): 303–11.
Withers, C. and Keighren, I. (2011) 'Travels into print: Authoring, editing and narratives of travel and exploration, c. 1815–c. 1857', *Transactions of the Institute of British Geographers* 36 (4): 560–73.
Woodward, R. (2004) *Military Geographies*. Oxford: Blackwell.
Wylie, J. (2005) 'A single day's walking: Narrating self and landscape on the South West Coast Path', *Transactions of the Institute of British Geographers* 30 (2): 234–47.
Wynne-Jones, S., North, P. and Routledge, P. (2015) 'Practising participatory geographies: Potentials, problems and politics', *Area* 47 (3). 218–21.
Young, I.M. (1990) *Throwing Like a Girl*. Bloomington, IN: Indiana University Press.
Yusoff, K. (2013) 'Geologic life: Prehistory, climate, futures in the Anthropocene', *Environment and Planning D: Society and Space*: 31 (5): 779–95.

INDEX